Organic synthesis
The roles of boron and silicon

Susan E. Thomas

OXFORD NEW YORK TOKYO
OXFORD UNIVERSITY PRESS
1991

Oxford University Press, Walton Street, Oxford OX2 6DP

Oxford New York Toronto
Delhi Bombay Calcutta Madras Karachi
Petaling Jaya Singapore Hong Kong Tokyo
Nairobi Dar es Salaam Cape Town
Melbourne Auckland

and associated companies in
Berlin Ibadan

Oxford is a trade mark of Oxford University Press

Published in the United States
by Oxford University Press, New York

A catalogue record for this book is available from the British Library

Library of Congress Cataloging in Publication Data
(Data available)
ISBN 0–19–855663–2
ISBN 0–19–855662–4 (Pbk)

Printed in Great Britain by
Information Press Ltd., Eynsham, Oxon.

Series Editor's Foreword

Modern synthetic organic chemistry allows the synthesis of virtually any desired complex molecular structure. Central to the synthetic chemist's armoury are reagents derived from boron and silicon, which can be used to effect a wide range of structural changes.

Oxford Chemistry Primers have been designed to provide concise introductions relevant to all students of chemistry, and contain only the essential material that would usually be covered in an 8–10 lecture course. In this first primer of the series Sue Thomas has produced an excellent account of two enormous topics which are described in a very easy to read and student-friendly fashion. This primer will be of interest to apprentice and master chemist alike.

S. G. D.

Preface

Boron and silicon compounds are well established in organic synthesis and a bewildering array of reactions involving these elements is reported every year. The small number of pages traditionally allotted to these elements in one-volume textbooks now fails to emphasize their importance and their wide range of uses.

This short text is intended to introduce the student of synthetic organic chemistry to the reactions of organoboron and organosilicon compounds which have been exploited by organic chemists, and to illustrate how these reactions have been applied to problems in organic synthesis. It is hoped that the chemistry described in this slim volume will encourage students to consult the more comprehensive reference texts and reviews available. These are listed in the bibliographies at the end of each section.

In view of the importance currently attached to the synthesis of homochiral organic molecules, examples which illustrate the use of organoboron and organosilicon compounds in this area are included where appropriate.

Finally, many thanks to Michael J. Harrison, M. Elena Lasterra–Sanchez, K. Gail Morris, Stephen P. Saberi, Matthew M. Salter, Gary J. Tustin, and K. Winky Young, who proof-read the manuscript.

S. E. T.

London
June 1991

Contents

B1   Hydroboration   1

B2   Reactions of organoboranes   9

B3   Further reactions of organoboranes   17

B4   Organoboron routes to unsaturated hydrocarbons   25

B5   Allylboranes and borane enolates   31

B6   Boronic ester homologation   42

Si1 Properties of organosilicon compounds   47

Si2 Protection of hydroxy groups as silyl ethers   51

Si3 Silyl enol ethers and relates silyl ethers   55

Si4 Alkene synthesis (Peterson olefination)   67

Si5 Alkynyl-, vinyl-, and arylsilanes   71

Si6 Allylsilanes and acylsilanes   84

Oxford Chemistry Primers

1   S. E. Thomas   *Organic synthesis: the roles of boron and silicon*

2   D. T. Davies   *Aromatic heterocyclic chemistry*

3   P. R. Jenkins   *Organometallic reagents in synthesis*

4   M. Sainsbury   *Aromatic chemistry*

5   L. M. Harwood   *Polar rearrangements*

6   I. E. Markó   *Oxidations*

B1. Hydroboration

B1.1 Characteristics of the hydroboration reaction

Hydroboration is the term given to the addition of a boron–hydrogen bond to either the carbon–carbon double bond of an alkene (Equation B1.1) or the carbon–carbon triple bond of an alkyne (Equation B1.2). The first examples of hydroboration were reported in 1956 from the laboratories of H.C. Brown and since then the reaction has found many important applications in organic chemistry. Indeed, in 1979 Brown was awarded the Nobel chemistry prize for his contributions to hydroboration and related areas of reactivity.

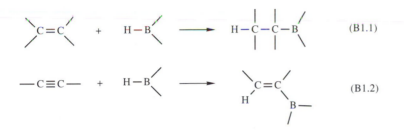

$$(B1.1)$$

$$(B1.2)$$

The simplest boron hydride is borane, BH_3, which dimerizes to diborane B_2H_6 in an equilibrium which lies overwhelmingly to the side of diborane under normal conditions of temperature and pressure (Equation B1.3).

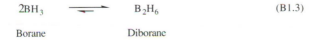

$$(B1.3)$$

Borane   Diborane

Diborane may be generated *in situ* from $NaBH_4$ and $AlCl_3$ or BF_3 but many sources of borane are commercially available, e.g. borane–tetrahydrofuran ($H_3B.THF$), borane–dimethyl sulphide ($H_3B.SMe_2$).

Unhindered alkenes react rapidly with borane to give initially monoalkylboranes, then dialkylboranes, and finally trialkylboranes. The reaction of borane with ethene is illustrated in Equation B1.4.

Equation B1.3 is represented below using structural formulae.

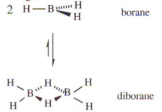

Note that in borane the hydrogen atoms form a trigonal planar arrangement around the boron atom whereas in diborane they are arranged tetrahedrally. The BHB bonds in diborane are examples of three-centre two-electron bonds.

The boron atom in BH_3 is sp^2 hybridized with a vacant *p* orbital perpendicular to the plane of the three boron–hydrogen bonds. Thus borane and its derivatives are electrophilic (Lewis acidic) and combine readily with electron-rich species. For example, borane interacts with one of the lone pairs on the oxygen atom of tetrahydrofuran as shown below.

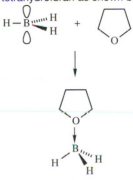

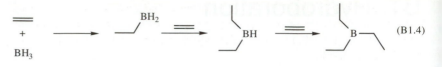

$$\text{(B1.4)}$$

often represented: often represented:

With increasing steric hindrance around the double bond, the second and third hydroboration steps become increasingly sluggish and so, for example, hydroboration of cyclohexene with H$_3$B.THF may be stopped at the dialkylborane stage, and hydroboration of 1,2-dimethylcyclopentene with H$_3$B.THF does not proceed beyond the monoalkylborane.

It has been observed that hydroboration exhibits the following characteristics:

a) **The boron atom adds preferentially to the least hindered end of an unsymmetrically substituted double bond** (Equations B1.5 and B1.6). This is consistent with the fact that boron is more positive than hydrogen (electronegativity of boron 2.01, electronegativity of hydrogen 2.20), but the regioselectivity is predominantly a result of steric factors rather than electronic factors.

$$\text{(B1.5)}$$

$$\text{(B1.6)}$$

b) **Controlled *cis*-addition of the boron–hydrogen bond to the alkene occurs** (Equation B1.7).

$$\text{(B1.7)}$$

c) **Addition of the boron–hydrogen bond to the carbon–carbon double bond takes place on the least hindered face** (Equation B1.8).

$$\text{(B1.8)}$$

The characteristic features of hydroboration are consistent with a concerted four-centre transition state carrying charges on the participating atoms (Figure B1.1) and this model adequately rationalizes the majority of hydroboration results.

Figure B1.1

It is of note that concerted addition of a boron–hydrogen bond to an alkene is not a forbidden reaction if the vacant orbital on boron is involved in the process.

The transition state shown in Figure B1.1 is thought to be preceded by a π-complex formed by donation of the π bond of the alkene into the vacant *p* orbital on boron.

B1.2 Alkylboranes as hydroborating reagents

Borane transforms a wide range of alkenes into trialkylboranes under mild conditions but the trifunctional nature of borane and its trialkylborane products imposes some limitations on its use. Many of the synthetically useful reactions of the trialkylboranes (see Chapters B.2 and B.3) use all three aklyl substituents, but some reactions only utilize either two or even one of the alkyl substituents. This sets a maximum yield (based on the alkene starting material) for these latter transformations of 66% and 33% respectively which is clearly undesirable especially if the alkene involved is the product of a multi-step synthetic sequence. To overcome this problem, and others such as the production of intractable polymers on addition of borane to dienes and alkynes, monoalkylborane and dialkylborane hydroborating reagents were introduced. Some commonly used reagents are depicted in Figure B1.2 and two are described in more detail below.

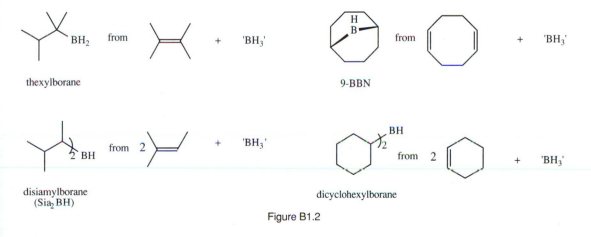

thexylborane

9-BBN

disiamylborane
(Sia$_2$BH)

dicyclohexylborane

Figure B1.2

Thexylborane

1,1,2-Trimethylpropylborane (thexylborane) is a monoalkylborane prepared by hydroboration of 2,3-dimethylbut-2-ene with $H_3B.THF$ (Equation B1.9). On standing at room temperature, the tertiary alkyl group slowly isomerizes to a primary alkyl group (see Section B3.1) and so the reagent is normally not stored but prepared and used as required.

(B1.9)

thexylborane

The presence of two boron–hydrogen bonds in thexylborane makes it ideal for the hydroboration of dienes. The reaction is much more reliable than the corresponding reaction using $H_3B.THF$ which generally tends to form polymeric organoboranes. As there is a strong preference for the formation of five- and seven-membered rings over six-membered and larger rings when the reaction is run under kinetic conditions, optimum yields are obtained when it is applied to 1,3- and 1,5-dienes (Equations B1.10 and B1.11). Hydroboration of dienes is often coupled with subsequent carbonylation and oxidation to give cyclic ketones (see Section B3.2).

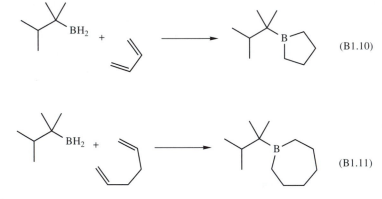

(B1.10)

(B1.11)

9-BBN

Addition of $H_3B.THF$ to 1,5-cyclooctadiene gives a mixture of 9-borabicyclo[4.2.1]nonane and 9-borabicyclo[3.3.1]nonane. On heating, the [4.2.1] system isomerizes to the thermodynamically more stable [3.3.1] compound which is known as 9-BBN (Equation B1.12). As 9-BBN is crystalline, relatively stable to air and heat, and is available from commercial sources, this dialkylborane is a popular hydroborating agent.

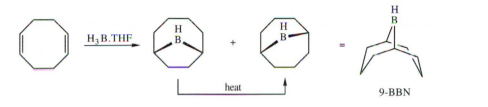

(B1.12)

9-BBN

sometimes represented:

Due to the considerably greater steric demands of 9-BBN, it is more regioselective in hydroboration reactions than borane as demonstrated by the examples given in Figure B1.3.

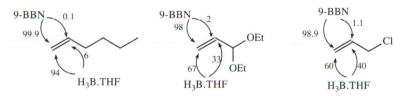

Figures represent % of product in which boron atom is found on carbon atom indicated
Figure B1.3

9-BBN hydroborates internal alkynes cleanly (Equation B1.13) and thus for this reaction it is superior to borane which tends to give intractable polymers when added to alkynes. The reaction is less useful for terminal alkynes as monohydroboration can only be achieved if an excess of alkyne is used.

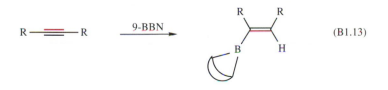

(B1.13)

B1.3 Alkylboranes used in asymmetric hydroboration

The hydroborating reagents described in Section B1.2 are generated from achiral alkenes. Addition of a source of borane to a homochiral alkene derived from nature's 'chiral pool' produces homochiral alkylboranes. A number of such reagents, which are used in asymmetric hydroboration reactions (see Section B1.4), are described below.

homochiral = enantiomerically pure

Dilongifolylborane (Lgf$_2$BH)

(+)-Longifolene (the world's most abundant sesquiterpene) is a substituted bicyclo[2.2.1]heptane system with an exocyclic double bond and a bridging hydrocarbon chain which very effectively shields the *exo* face of the double bond. Thus, in contrast to the behaviour of norbornene which is hydroborated on its *exo* face (Equation B1.8), (+)-longifolene is hydroborated on its *endo* face with the boron adding to the least hindered end of the double bond to give the reagent dilongifolylborane (Lgf$_2$BH) (Equation B1.14).

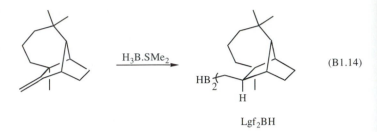

(B1.14)

Lgf$_2$BH

Lgf$_2$BH is a stable crystalline solid of limited solubility in most solvents used for hydroboration (e.g. THF, diethyl ether, hexane). Thus disappearance of the solid is a useful indicator of the progress of a hydroboration reaction performed using this reagent.

Diisopinocampheylborane (Ipc$_2$BH) and monoisopino-campheylborane (IpcBH$_2$)

Hydroboration of α-pinene gives diisopinocampheylborane (Ipc$_2$BH) or monoisopinocampheylborane (IpcBH$_2$) depending on the reaction conditions used. In contrast to longifolene, both enantiomers of α-pinene are readily available and so both enantiomers of Ipc$_2$BH and IpcBH$_2$ are accessible (Figure B1.4). Note that hydroboration occurs on the least hindered face of α-pinene, i.e. the face not obstructed by the dimethyl bridge, and the boron atom adds to the least substituted end of the alkene.

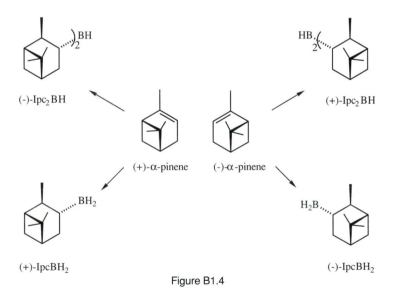

Figure B1.4

B1.4 Asymmetric hydroboration

Consider hydroboration of the prochiral alkene 2-methylbut-1-ene by the homochiral hydroborating reagent (+)-Ipc$_2$BH (Figure B1.5).

2-Methylbut-1-ene is prochiral because the products formed after addition of reagents to its double bond are chiral.

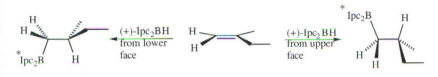

* fragment contains homochiral centres

Figure B1.5

The hydroborating reagent may approach either face of the alkene in order that the boron–hydrogen bond and the carbon–carbon double bond may interact in the hydroboration reaction. As the boron–hydrogen bond approaches the alkene, interactions between the substituents on the borane and the substituents on the alkene become important. When a homochiral hydroborating reagent is used, the interactions which arise as it approaches one face of the alkene differ from the interactions which arise as it approaches the other face of the alkene. This is a result of the chiral centres in the hydroborating reagent. The approach which leads to the least unfavourable interactions, i.e. the approach which involves the lowest energy transition state, is favoured and the two possible diastereoisomeric hydroboration products are formed in unequal amounts. This is known as asymmetric hydroboration. (When an achiral hydroborating reagent is used, approach from either face is equally probable as the interactions which arise between the hydroborating reagent and the alkene are energetically equivalent for either trajectory.)

The efficiency of asymmetric hydroboration is high if one approach trajectory leads to severe steric interactions between the hydroborating reagent and the alkene and the approach trajectory to the other face of the alkene involves relatively insignificant steric interactions, i.e. the energy difference between the two transition states is large. It should be noted, however, that if both approaches involve major steric interactions then a decrease in overall reactivity will be observed.

Subjecting boranes produced by asymmetric hydroboration to further reactions such as oxidation (see Section B2.1) leads to optically active products. For example, oxidation of the products of the reaction depicted in Figure B1.5 gives (R)- and (S)-2-methylbutan-1-ol in 21% e.e. in favour of the (R) enantiomer (Figure B1.6). (Note that this result reveals that the (+)-Ipc$_2$BH preferentially attacks the upper face of 2-methylbut-1-ene.)

If the face discrimination in the asymmetric hydroboration reaction is high then the optical purity of the chiral molecule produced will also be high. Efficient asymmetric hydroboration reactions followed by stereospecific cleavage of the boron–carbon bonds produced have been used in syntheses of several complex homochiral molecules (see Section B2.1).

The two transition states for the addition of a homochiral hydroborating reagent to the two faces of a prochiral alkene are diastereoisomeric and of different energy.

The two transition states for the addition of an achiral hydroborating reagent to the two faces of a prochiral alkene are enantiomeric and of equal energy.

e.e. = enantiomeric excess

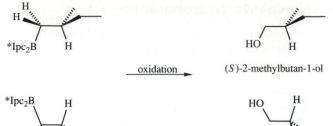

oxidation

(*S*)-2-methylbutan-1-ol

(*R*)-2-methylbutan-1-ol (21% e.e.)

* fragment contains homochiral centres

Figure B1.6

B2. Reactions of organoboranes

The empty *p* orbital on the boron atom of organoboranes renders them electrophilic and highly susceptible to attack by nucleophiles. The tetrahedral species so formed is known as an organoborate (Equation B2.1).

$$(B2.1)$$

organoborane   nucleophile   organoborate

If the nucleophile bears a leaving group (or an alternative electron sink) then 1,2-migrations occur very easily (Equation B2.2). Note carefully that in the migration step the migrating alkyl group takes with it both the electrons from its bond to boron (thus rendering the boron atom in the product neutral), and that the migrating alkyl group and the leaving group are antiperiplanar to each other.

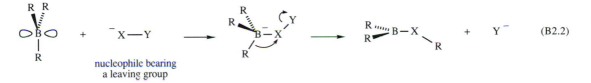

$$(B2.2)$$

nucleophile bearing
a leaving group

In the migration step negative charge builds up on the migrating group and this is reflected in the relative migratory aptitudes of alkyl groups which is primary > secondary > tertiary. Not all reactions follow this pattern, however, and relative migratory aptitudes depend on other factors such as steric and conformational effects.

As will be seen below and in following chapters, attack by nucleophiles and subsequent 1,2-migration reactions dominate much of the reactivity of organoboranes.

B2.1 Oxidation

Organoboranes are normally handled under a nitrogen atmosphere as they are generally sensitive to oxidation processes. When oxidation is actually required, it is most commonly carried out using alkaline hydrogen peroxide although many other oxidizing systems have been used, including several chromium reagents.

Boron–oxygen bond strengths (480–565 kJ mol^{-1}) are greater than boron–carbon bond strengths (350–400 kJ mol^{-1}). This reflects an interaction between the empty *p* orbital on boron and an electron pair in one of the oxygen's two filled *p* orbitals.)

Oxidation using alkaline hydrogen peroxide

Oxidation of alkylboranes by alkaline hydrogen peroxide produces alcohols. The reaction is essentially quantitative and has been successfully applied to a wide variety of alkylboranes (Equations B2.3–5). It is important to note that the stereochemistry of the carbon atom attached to the boron atom is retained in this conversion of a carbon–boron bond to a carbon–oxygen bond (Equation B2.5).

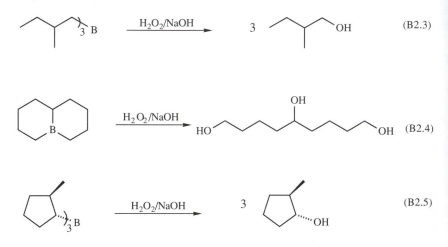

On combination with alkene hydroboration, the resulting two-step process is a very important, widely-used transformation which may be regarded as *anti*-Markovnikov hydration of the alkene (Equation B2.6).

The mechanism of borane oxidation by alkaline hydrogen peroxide is depicted in Figure B2.1. Due to its empty $2p$ orbital, the boron atom of the trigonal planar trialkylborane is electrophilic and is attacked by the hydroperoxide anion to give a tetrahedral borate anion in step 1. In step 2 an alkyl group migrates from boron to oxygen to liberate hydroxide ion and form a stable boron–oxygen bond. Note that this step occurs with retention of configuration at the migrating carbon atom. Repetition of steps 1 and 2 transfers the remaining alkyl groups from boron to oxygen to give a trialkoxyborane. Finally, hydrolysis of the carbon–oxygen bonds of the trialkoxyborane gives three molecules of alcohol and one equivalent of sodium borate.

The alkyl group migrates with the two electrons from its bond to boron and as a result the migration occurs with retention of the stereochemistry of the alkyl group.

Oxidation of alkenylboranes by alkaline hydrogen peroxide gives aldehydes or ketones depending on the substituent pattern of the alkenyl group; thus, when alkaline hydrogen peroxide oxidation is combined with alkyne hydroboration, the resulting two-step process is a procedure for converting alkynes to carbonyl compounds (Equations B2.7 and B2.8).

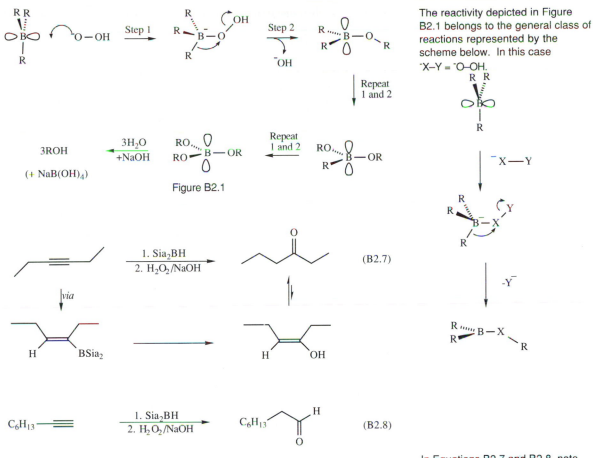

The reactivity depicted in Figure B2.1 belongs to the general class of reactions represented by the scheme below. In this case $^-X–Y = {}^-O–OH$.

Figure B2.1

$$3ROH \quad \xleftarrow[+NaOH]{3H_2O}$$

$(+ NaB(OH)_4)$

1. Sia$_2$BH
2. H$_2$O$_2$/NaOH

(B2.7)

|via|

1. Sia$_2$BH
2. H$_2$O$_2$/NaOH

C_6H_{13} ——— (B2.8)

In Equations B2.7 and B2.8 note that an alkenyl group migrates in preference to a secondary alkyl group.

   Asymmetric hydroboration followed by oxidation is used to give optically active alcohols. For example, addition of (+)-IpcBH$_2$ to 1-phenylcyclopentene followed by oxidation gives (1S,2R)-*trans*-2-phenylcyclopentanol in 100% e.e. (Equation B2.9). The structure of the product alcohol reveals that the homochiral hydroborating reagent encounters fewer unfavourable steric interactions with alkene substituents if it approaches the lower face of the alkene as drawn in Equation B2.9. This preference determines the absolute stereochemistry of the product. (The regiochemistry and relative stereochemistry of the product are determined by fundamental hydroboration characteristics.)

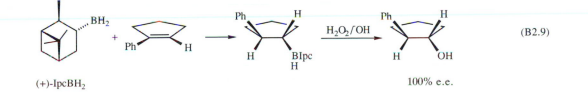

(+)-IpcBH$_2$

H$_2$O$_2$/OH

100% e.e.

(B2.9)

Homochiral alcohols produced by asymmetric hydroboration/oxidation have been used in syntheses of complex homochiral organic molecules such as (3R,3'R)-Zeaxanthin, a yellow pigment found in such diverse products as maize, egg yolk, and adipose tissue, and its enantiomer (3S,3'S)-Zeaxanthin (Figure B2.2). An achiral intermediate, derived from safranal, is asymmetrically hydroborated either by (+)-Ipc$_2$BH or by (−)-Ipc$_2$BH to give alkylboranes which are then oxidized and acidified to give homochiral intermediates. The intermediates which contain all the chirality present in the target molecules are then transformed by conventional steps into the two enantiomeric Zeaxanthins.

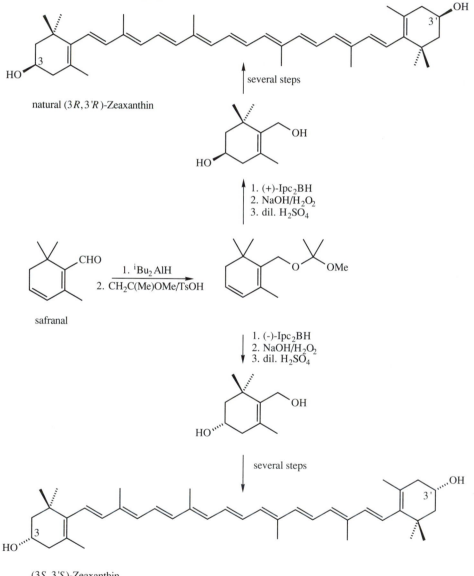

Figure B2.2

Oxidation using chromic acid

Aqueous chromic acid has been used to oxidize alkylboranes derived from cyclic alkenes to ketones. For example, hydroboration and oxidation of 1-methylcyclohexene converts it into 2-methylcyclohexanone (Equation B2.10).

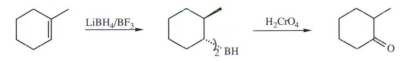

$$\text{(B2.10)}$$

B2.2 Protonolysis

Alkylboranes are readily protonolysed by carboxylic acids (but not by water, aqueous mineral acid, or aqueous alkali). The reaction is normally carried out by heating the alkylborane with excess propanoic acid or ethanoic acid in diglyme (Equation B2.11).

$$R_3B \xrightarrow[\text{diglyme, 165 °C}]{\text{excess } C_2H_5CO_2H} 3 \text{ RH} \qquad \text{(B2.11)}$$

Protonolysis proceeds with retention of configuration of the alkyl group as depicted in Equation B2.12. Note also that the overall effect of hydroboration–protonolysis is *cis* addition of hydrogen to the alkene. Thus the two-reaction sequence provides an alternative to catalytic hydrogenation which is useful in cases where catalytic hydrogenation fails, e.g. hydrogenation of carbon–carbon double bonds in molecules containing sulphur groups.

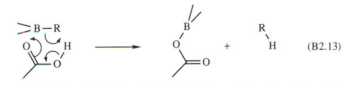

$$\text{(B2.12)}$$

Retention of configuration of the alkyl group is consistent with the *concerted* cyclic mechanism used to explain why carboxylic acids alone protolytically cleave carbon–boron bonds (Equation B2.13).

$$\text{(B2.13)}$$

Protonolysis of alkenylboranes by carboxylic acids occurs readily. The stereochemistry of the alkenyl group is retained during the reaction and so hydroboration/protolytic cleavage of alkynes leads to *cis* alkenes. Deuterated

Catecholborane is formed by
addition of catechol to H$_3$B.THF.

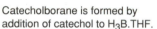

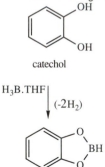

catechol

H$_3$B.THF $\Big\downarrow$ (-2H$_2$)

catecholborane
Due to its boron–oxgen bonds it is a
less reactive hydroborating reagent
than H$_3$B, H$_2$BR, or HBR$_2$. It is often
used for hydroboration of alkynes.

boranes and carboxylic acids can be used to synthesize specifically labelled
alkenes as shown in Figure B2.3. (B$_2$D$_6$ may be generated from LiAlD$_4$ and
BF$_3$.)

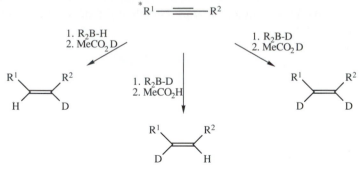

*R^1 bulkier than R^2

Figure B2.3

For example, 3,3-dimethylbutyne has been reductively deuterated in a
controlled manner by treatment with catecholborane followed by deuterated
ethanoic acid (Equation B2.14).

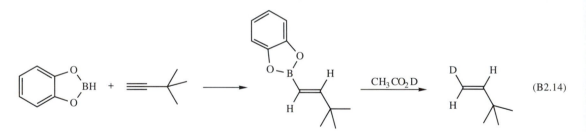

(B2.14)

B2.3 Halogenolysis

Cleavage of boron–carbon bonds by halogens does not proceed efficiently.
The elements of HI, HBr, and HCl can, however, be added to alkenes in an
anti-Markovnikov fashion by hydroboration and subsequent addition of either
I$_2$/NaOMe, Br$_2$/NaOMe, or NCl$_3$ (Equations B2.15–17).

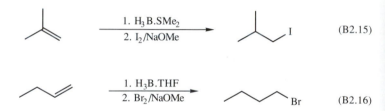

(B2.15)

(B2.16)

(B2.17)

The iodination and bromination of intermediate alkylboranes proceed with clean inversion of configuration whilst the radical chlorination reaction leads to loss of stereochemistry (Figure B2.4).

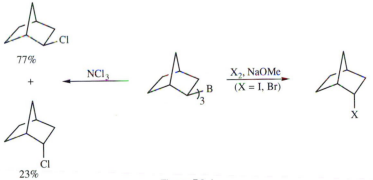

Figure B2.4

A mechanism consistent with the observed characteristics of the iodination and bromination reactions has been proposed and is illustrated in Figure B2.5 for the iodination reaction.

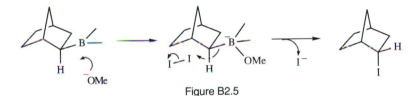

Figure B2.5

B2.4 Amination

Alkylboranes are converted to primary amines by amines bearing good leaving groups such as chloroamine or *O*-hydroxylaminesulphonic acid (Equation B2.18).

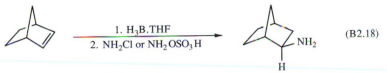

(B2.18)

The reaction proceeds with retention of stereochemistry *via* the mechanism illustrated in Figure B2.6 and the overall transformation may be regarded as the *cis* addition of ammonia across a carbon–carbon double bond. (Note that the third alkyl group on boron does not participate in the reaction, thus reducing the maximum yield to 67%.)

The reactivity depicted in Figure B2.6 belongs to the general class of reactions represented by the scheme below. In this case ⁻X–Y = H_2N–Cl or H_2N–OSO_3H.

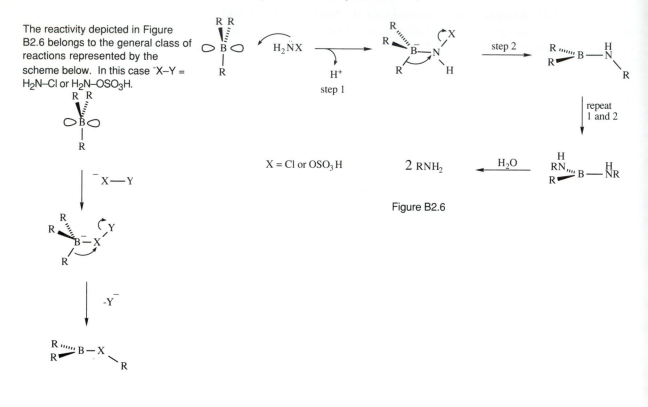

Figure B2.6

B3. Further reactions of organo-boranes

B3.1 Isomerization

On heating, alkylboranes isomerize and the boron atom moves to a position where steric interactions are minimized (Equations B3.1 and B3.2).

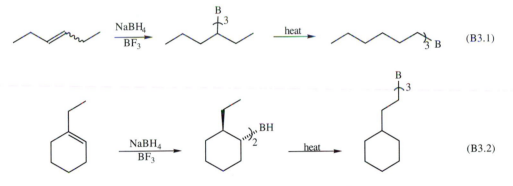

(B3.1)

(B3.2)

Heating an alkylborane facilitates elimination of the boron and an α hydrogen atom to give a boron–hydrogen bond and an alkene. Readdition of the boron–hydrogen group to the alkene gives either the original alkylborane or an isomeric alkylborane depending on the orientation of addition. Thus an equilibrium is set up which is driven towards the most stable alkylborane. Equation B3.3 illustrates the individual equilibria involved in the example depicted in Equation B3.1.

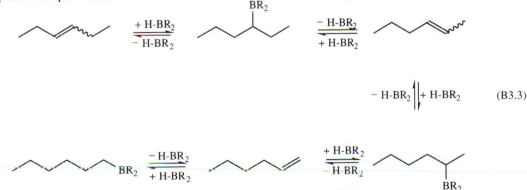

(B3.3)

Elimination/readdition equilibria are the basis of a solution to the problem of converting thermodynamically more stable internal alkenes into thermodynamically less stable terminal alkenes. For example, the trisubstituted alkene 3-ethylpent-2-ene is converted to the monosubstituted alkene 3-ethylpent-1-ene by a hydroboration/isomerization/*displacement* sequence (Equation B3.4).

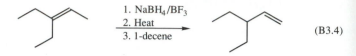

1. NaBH$_4$/BF$_3$
2. Heat
3. 1-decene

(B3.4)

The product of hydroboration is heated to convert it to its least sterically crowded isomer (Equation B3.5), and then 1-decene is added. This sets up the overall elimination/readdition equilibrium shown in Equation B3.6 from which the more volatile product alkene is distilled.

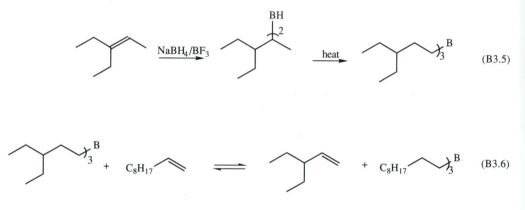

(B3.5)

(B3.6)

B3.2 Carbonylation

Carbonylation reactions of alkylboranes are some of the most widely applicable and synthetically useful reactions of these molecules. Carbonylation transforms alkylboranes into many products including aldehydes, ketones, and tertiary alcohols.

These transformations share common initial steps in which the carbon monoxide interacts with the trialkylborane to give an intermediate organoborate. This readily transfers one of its alkyl groups to the carbon atom derived from carbon monoxide to give intermediate **X** (Figure B3.1).

Note that the reactivity depicted in Figure B3.1 falls into the general class of reaction illustrated in Equation B2.2.

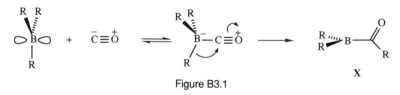

Figure B3.1

Carbonylation leading to aldehydes

If carbonylation of a trialkylborane is performed in the presence of a metal hydride such as LiAlH(OMe)$_3$, then intermediate **X** is reduced. Subsequent oxidation by alkaline hydrogen peroxide (see Section B2.1) gives an aldehyde product (Figure B3.2).

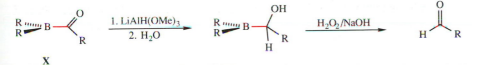

Figure B3.2

Thus hydroboration followed by carbonylation in the presence of a metal hydride may be used to hydroformylate an alkene as exemplified by the reaction sequence shown in Equation B3.7.

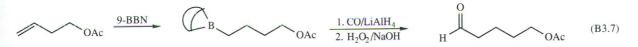

(B3.7)

Carbonylation leading to ketones

If carbonylation of a trialkylborane is conducted in the presence of water, the water promotes migration of a second alkyl group from the boron centre to the carbon atom derived from carbon monoxide. Subsequent oxidation by alkaline hydrogen peroxide gives a ketone which bears two substituents derived from the trialkylborane. A pathway which accounts for this is shown in Figure B3.3.

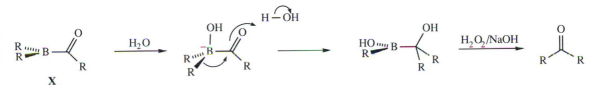

Figure B3.3

This reaction has been developed into a versatile synthesis of ketones outlined in Equation B3.8. It is based on thexylborane as the thexyl group shows a very low tendency to migrate in the carbonylation reaction, and so the alkyl groups derived from alkenes x and y are transferred efficiently from boron to carbon.

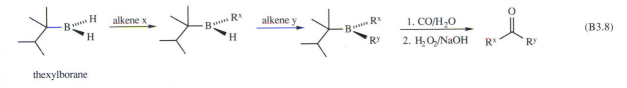

(B3.8)

thexylborane

Examples of applications of this strategy are given below.
a) The unsymmetrical ketone juvabione, a molecule which possesses high juvenile hormone activity, has been synthesized from two readily-available alkenes (Equation B3.9). (The chiral centre present in the first alkene that reacts with thexylborane does not exert any control over which face of the

adjacent alkene reacts with the thexylborane and so juvabione is formed as a diastereoisomeric mixture.)

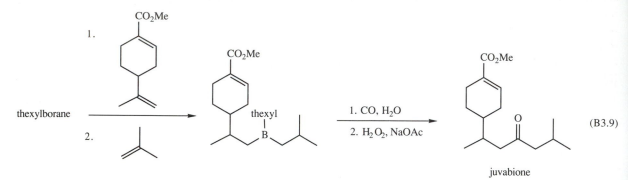

(B3.9)

juvabione

b) If the two susbstrate alkenes are in the same molecule, then ketone formation generates a cyclic product. Equation B3.10 illustrates a synthesis of the thermodynamically disfavoured *trans*-perhydroindan-1-one in which the stereochemistry is introduced cleanly into the product by (i) stereospecific *cis* hydroboration and (ii) retention of stereochemistry as the chiral alkyl group migrates from boron to carbon in the carbonylation step.

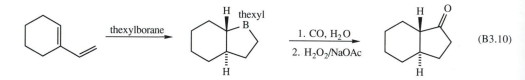

(B3.10)

c) A similar reaction sequence has been used in the construction of a steroid skeleton (Equation B3.11). Note again that stereospecific *cis* hydroboration of the trisubstituted alkene and retention of stereo-chemistry as the chiral group attached to boron migrates to the carbon atom derived from carbon monoxide result in excellent stereochemical control in the hydroboration/carbonylation/oxidation sequence.

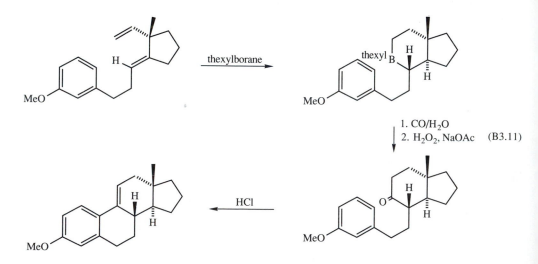

(B3.11)

Carbonylation leading to tertiary alcohols

Carbonylation of a trialkylborane in the presence of ethylene glycol promotes migration of both the second and third alkyl groups from the boron atom of intermediate **X** to the carbon atom derived from carbon monoxide. Subsequent oxidation by hydrogen peroxide in this case produces a tertiary alcohol which bears three substituents derived from the trialkylborane (Figure B3.4).

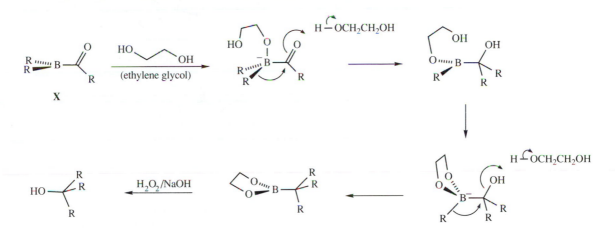

Figure B3.4

Thus carbonylation of a trialkylborane in the presence of ethylene glycol results in the boron atom of the trialkylborane being replaced by a C–OH unit (Equation B3.12).

$$R_3B \quad \xrightarrow[\text{2. } H_2O_2/\text{NaOH}]{\text{1. CO/HOCH}_2\text{CH}_2\text{OH}} \quad R_3COH \qquad \text{(B3.12)}$$

This forms the basis of an attractive synthetic method for converting polyenes into carbocyclic structures (Equation B3.13).

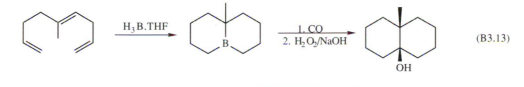

(B3.13)

B3.3 Cyanidation

Addition of the cyanide anion (which is isoelectronic with carbon monoxide) to alkylboranes produces stable organoborates. Treatment of these with electrophiles such as trifluoroacetic anhydride induces alkyl-group migration.

Note that the three alkyl-group migrations which occur in Figure B3.5 fall into the general class of reaction illustrated in Equation B2.2.

Below room temperature, two alkyl groups are transferred and oxidation gives ketones. If an excess of trifluoroacetic anhydride and higher temperatures are used then the third alkyl group can be induced to migrate and oxidation leads to tertiary alcohols (figure B3.5).

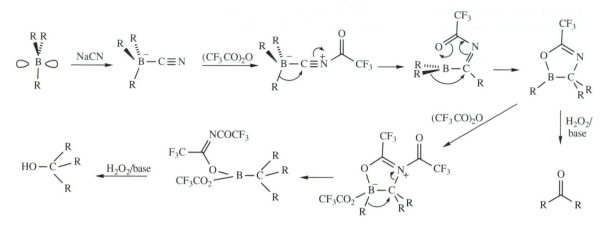

Figure B3.5

The reaction, known as the *cyanoborate process*, is experimentally simpler and occurs under milder conditions than carbonylation. It has been used, for example, in the synthesis of a tertiary alcohol precursor to a tridentate metal ligand (Equation B3.14). The alternative carbonylation route to this compound resulted in extensive dehydrobromination due to the higher temperature required.

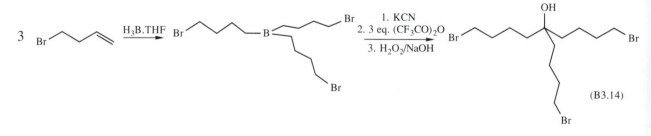

(B3.14)

B3.4 Reaction with dichloromethyl methyl ether

Note that the three migrations of alkyl groups from boron to carbon which occur in Equation B3.15 fall into the general class of reactivity represented in Equation B2.2.

Trialkylboranes react rapidly with dichloromethyl methyl ether in the presence of a hindered base. Transfer of all three alkyl groups from the boron atom of the intermediate organoborate occurs and subsequent oxidation produces the tertiary alcohol derived from the three alkyl groups of the alkylborane (Equation B3.15).

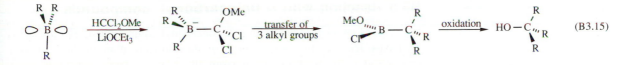

(B3.15)

Tertiary alkyl groups cannot be transferred readily in the carbonylation reaction or the cyanoborate process. In contrast, transfer of tertiary alkyl groups occurs smoothly in the dichloromethyl methyl ether reaction (Equation B3.16).

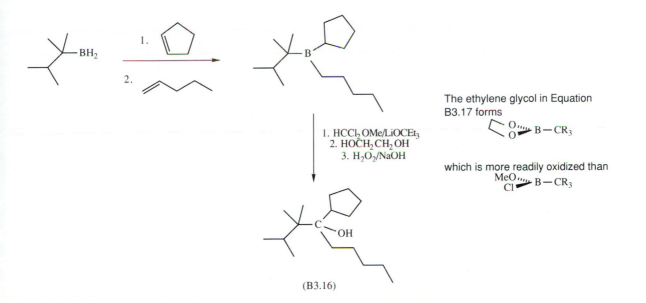

The ethylene glycol in Equation B3.17 forms

which is more readily oxidized than

(B3.16)

Migration of tertiary alkyl groups in the dichloromethyl methyl ether reaction does however mean that ketones cannot be synthesized using thexylborane and the approach adopted in the carbonylation and cyanidation processes. In order to obtain ketones by this route, boranes bearing chloro or alkoxy groups, i.e. groups of very low migratory aptitude, have to be used as substrates (Equation B3.17).

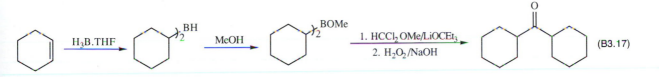

(B3.17)

B3.5 Reaction with α-halocarbonyl compounds

Reaction of an alkylborane with α-halocarbonyl compounds in the presence of a base leads to the replacement of the halogen atom in the carbonyl compound with an alkyl group from the borane. The reaction is most commonly applied to α-bromoesters (Equation B3.18).

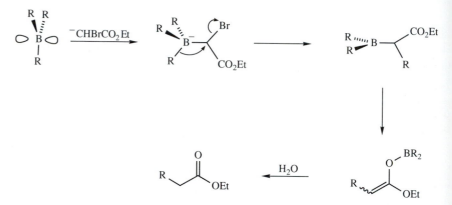

$$(B3.18)$$

Addition of the anion of the α-bromoester to the alkylborane initiates the reaction. Transfer of one alkyl group then takes place and the resulting α-boryl ester tautomerizes to an alkenyloxyborane. Finally hydrolysis releases the alkylated ester (Figure B3.6).

Note that the alkyl-group migration from boron to carbon which occurs in Figure B3.6 belongs to the general class of reaction depicted by Equation B2.2.

Figure B3.6

As only one alkyl group migrates in the reaction, 9-BBN derivatives are often used to prevent wastage of a more valuable alkyl group. (The alkyl groups of the bicyclononane system show a very low migratory aptitude.) Note that, as observed in many related migrations, the stereochemistry of the migrating alkyl group is retained (Equation B3.19).

$$(B3.19)$$

B4. Organoboron routes to unsaturated hydrocarbons

B4.1 Synthesis of alkenes

(*E*)-Alkenes

Hydroboration of 1-haloalk-l-ynes followed by treatment with sodium methoxide and then acetic acid produces (*E*)-alkenes (Equation B4.1).

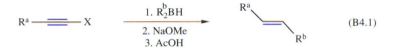

$$R^a \!\!\!=\!\!\! X \quad \xrightarrow[\substack{2.\ NaOMe \\ 3.\ AcOH}]{1.\ R^b_2BH} \quad R^a\!\!\diagdown\!\!=\!\!\diagup\!\!R^b \qquad (B4.1)$$

1-Haloalk-l-ynes may be generated by halogenation of the corresponding lithium acetylides at low temperature.

$$R^a\!\!\!=\!\!\!H$$
$$\downarrow RLi$$
$$R^a\!\!\!=\!\!\!Li$$
$$\downarrow X_2$$
$$R^a\!\!\!=\!\!\!X$$

The pathway followed by the reaction is depicted in Figure B4.1. Methoxide anion adds to the α-haloalkenylborane generated by hydroboration of the haloalkyne, and induces migration of an alkyl group from the boron atom to the alkenyl carbon atom. The migration displaces halide anion from the alkenyl carbon atom and the centre is inverted. Finally protonolysis of the carbon–boron bond by acetic acid releases the (*E*)-alkene.

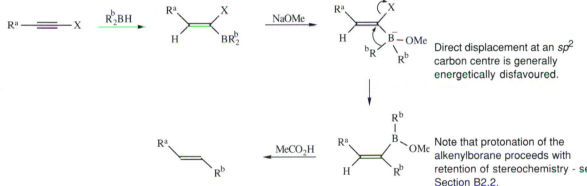

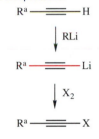

Direct displacement at an *sp²* carbon centre is generally energetically disfavoured.

Note that protonation of the alkenylborane proceeds with retention of stereochemistry - see Section B2.2.

Figure B4.1

Only one alkyl group is transferred from boron to carbon in the reaction and so generation of the required dialkylborane from thexylborane prevents wastage of more valuable alkyl groups (Equation B4.2). Note that the migrating alkyl group involved in the reaction sequence depicted in Equation B4.2 retains its stereochemistry.

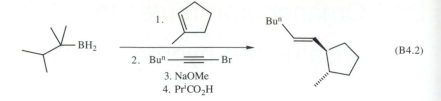

$$(B4.2)$$

(Z)-Alkenes

Hydroboration of alk-l-ynes followed by addition of sodium hydroxide and iodine gives (Z)-alkenes (Equation B4.3).

$$R^a \equiv\!\!\!\equiv\!\!\!\equiv H \quad \xrightarrow[\text{2. NaOH/I}_2]{\text{1. R}^b_2 BH} \quad \overset{R^a}{\diagdown}\!\!=\!\!\overset{R^b}{\diagup} \qquad (B4.3)$$

The reaction is thought to proceed as shown in Figure B4.2. After hydroboration and addition of hydroxide anion to the boron atom, an iodonium species is formed which creates an electrophilic terminus for alkyl migration. (The migration occurs with retention of configuration of the alkyl group.) Finally an antiperiplanar elimination reaction gives the (Z)-alkene.

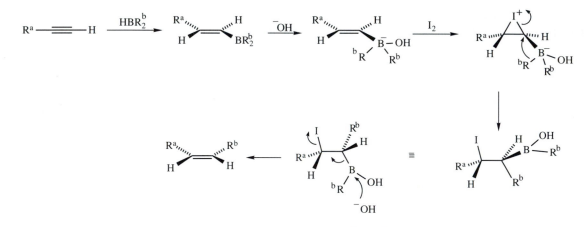

Figure B4.2

The reaction utilizes only one of the alkyl groups of the dialkylborane. Unfortunately, thexyl groups have been observed to migrate in this reaction and so the use of thexylborane as substrate is not so advantageous as in previously described reactions. Nevertheless, the reaction may be used with readily-available alkenes (Equation B4.4).

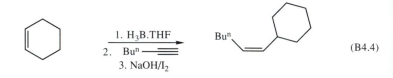

$$(B4.4)$$

Bromoborane derivatives have been used to circumvent the migration problem. Figure B4.3 depicts a synthesis of muscalure, a pheromone of the housefly *Musca domestica*, which uses dibromoborane as its boron source. In this synthesis, the single boron–hydrogen bond present in dibromoborane is used to hydroborate tridec-1-ene, and then $LiAlH_4$ reduction generates a new boron–hydrogen bond which in turn is used to hydroborate dec-1-yne. On iodination of the bromoborane produced, only the alkyl chain migrates and muscalure is produced in good yield.

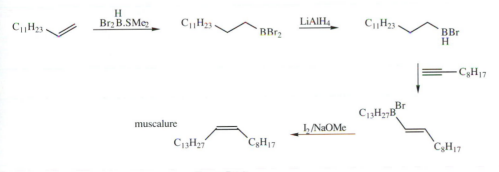

Figure B4.3

B4.2 Synthesis of alkynes, diynes, and enynes

Alkynes

Addition of the anion generated from a monosubstituted alkyne (or ethyne) to an alkylborane followed by iodination places an alkyl group from the alkylborane on the alkyne (Equation B4.5).

$$R^a \!\!-\!\!\!\equiv \quad \xrightarrow[\substack{2.\ R^b_3B \\ 3.\ I_2}]{1.\ Bu^nLi} \quad R^a \!\!-\!\!\!\equiv\!\!-\! R^b \qquad \text{(B4.5)}$$

After addition of the alkyne anion to the alkylborane, iodination facilitates alkyl group migration from boron to carbon in a transfer that resembles the one seen in the synthesis of (Z)-alkenes described in Section B4.1. Elimination to give the product alkyne occurs under the iodination reaction conditions (Figure B4.4).

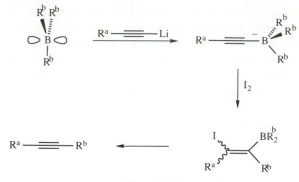

Figure B4.4

Primary alkyl, secondary alkyl, and aryl groups all migrate readily, and migration occurs with retention of configuration. The reaction is thus more versatile than the deprotonation/alkylation approach to substituted alkynes, which is generally only efficient for primary electrophiles and does not proceed at all for aryl halides. For example, triphenylborane may be used to incorporate a phenyl group into an alkyne (Equation B4.6).

$$PhMgBr \quad + \quad F_3B.OEt_2$$

$$\downarrow$$

$$Ph_3B$$

$$Ph_3B \quad \xrightarrow[\text{2. } I_2]{\text{1. } Bu^t \!\!=\!\! Li} \quad Bu^t \!\!=\!\! Ph \qquad (B4.6)$$

Diynes and enynes

Disiamylborane may be converted into dialkynyldisiamylborates or alkenylalkynyldisiamylborates by the sequences shown in Figures B4.5 and B4.6. Iodination of these compounds initiates migration of an alkynyl or alkenyl group to give diynes or enynes respectively by pathways resembling the one depicted in Figure B4.4. (Note that the (*E*)-geometry of the alkenyl group is retained in the latter reaction revealing that the iodine attacks the alkynyl group and not the alkenyl group.)

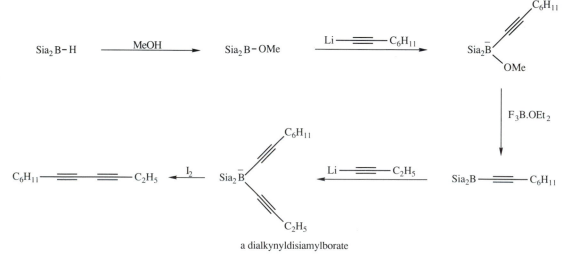

a dialkynyldisiamylborate

Figure B4.5

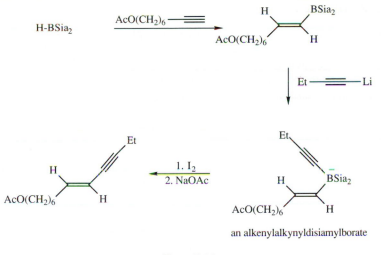

Figure B4.6

B4.3 Synthesis of dienes

Combination of reactions seen in Sections B4.1 and B4.2 with hydroboration/protonolysis reactions provides syntheses of stereodefined dienes.

(*E*, *E*)-Dienes

Incorporation of an alkenyl group into the (*E*)-alkene synthesis and its subsequent migration followed by a protonolysis step gives (*E*, *E*)-dienes (Figure B4.7).

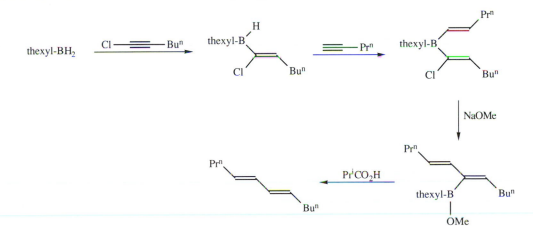

Figure B4.7

(E, Z)-Dienes

Hydroboration/protonolysis of (E)-enynes completes a route to (E, Z)-dienes. For example, the (E)-enyne generated in Figure B4.6 was converted into (7E, 9Z)-7,9-dodecadien-1-yl acetate (a natural sex pheromone of the European grape vine moth *Lobesia botrana*) by hydroboration with Sia_2BH followed by acetic acid protonolysis (Equation B4.7).

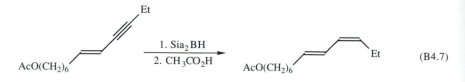

$$\text{(B4.7)}$$

(Z, Z)-Dienes

Hydroboration/protonolysis of diynes may be used to generate (Z, Z)-dienes (Equation B4.8).

$$\text{(B4.8)}$$

B5. Allylboranes and boron enolates

B5.1 Allylboranes in organic synthesis

Stability of allylboranes

Allylboranes rearrange at room temperature. For example, tricrotylborane undergoes the equilibria shown in Figure B5.1, and so attempts to prepare either the (E)-isomer or the (Z)-isomer of tricrotylborane at room temperature give the equilibrium mixture shown $((E):(Z) = 7:3)$ rather than isomerically pure products.

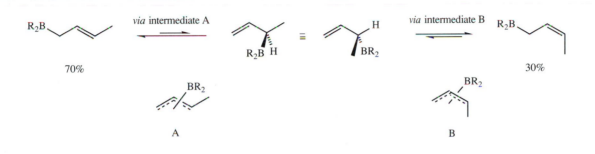

Figure B5.1

Allyldialkylboranes can, however, be used in further reactions (see below) if either (a) they are prepared and used directly at low temperature, or (b) the predominant isomer at equilibrium is the required isomer.

The rearrangement process depicted in Figure B5.1 involves interaction of the vacant p orbital on boron with the alkene. The presence of π-donor substituents on boron such as $-OR$ or NR_2 reduces the electron deficiency on boron and suppresses the rearrangement. Thus allylboron derivatives with two oxygen substituents, for example, are stable at room temperature and their (E)- and (Z)-isomers can be prepared isomerically pure (see below).

Reaction of allylboranes with carbonyl compounds

Triallylboranes react with aldehydes and ketones to give, on hydrolysis, homoallylic alcohols. The reaction proceeds stepwise through chair-like transition states (Figure B5.2).

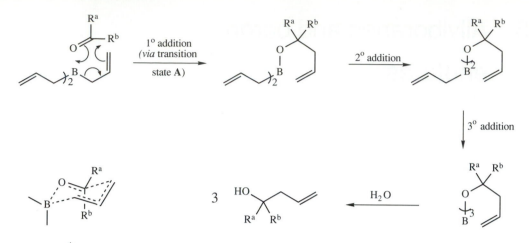

A

Figure B5.2

B-Allyl-9-BBN is prepared by the addition of an aluminium derivative of allyl bromide to B-methoxy-9-BBN.

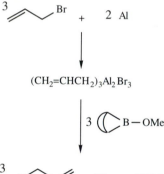

B-allyl-9-BBN

Although the first two additions proceed under relatively mild conditions, temperatures of over 100 °C are required to effect the third addition, and so 9-BBN derivatives often prove convenient substrates for the reaction. An example of the use of *B*-allyl-9-BBN is depicted in Equation B5.1.

(B5.1)

B-allyl-9-BBN

Synthesis of homochiral homoallylic alcohols containing one chiral centre

Addition of certain homochiral allylboranes to simple prochiral aldehydes produces homochiral homoallylic alcohols. For example, the allyldialkylborane derived from (+)-α-pinene and allylmagnesium bromide adds to benzaldehyde to give the homoallylic alcohol product in 96% e.e. (Figure B5.3).

Similarly a homochiral allyldialkoxyborane derived from (+)-camphor and allylmagnesium bromide adds to ethanal to give product homoallylic alcohol in 86% e.e. (Figure B5.4).

3,3-Dimethylallyldiisopinocampheylborane, obtained by hydroboration of 3-methylbuta-1,2-diene with (-)-Ipc$_2$BH, has been used to synthesize the 'irregular' monterpene (+)-artemisia alcohol in 95% e.e. (Figure B5.5).

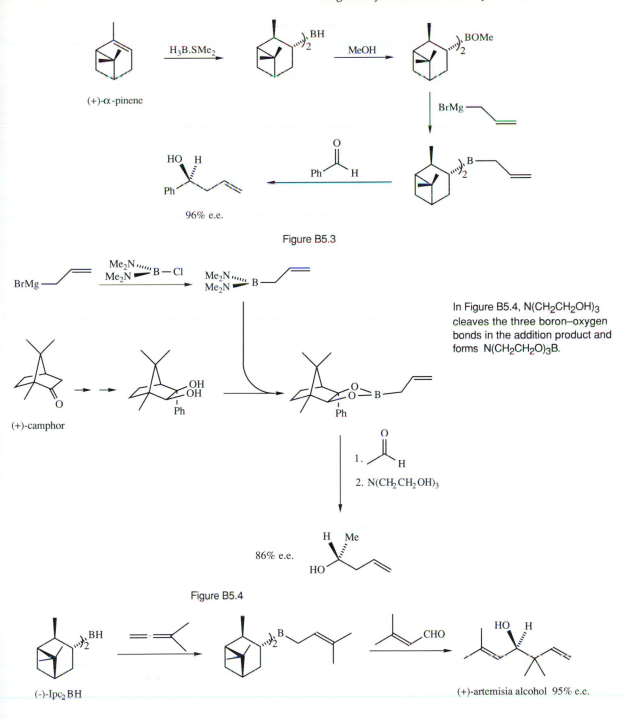

Figure B5.3

In Figure B5.4, $N(CH_2CH_2OH)_3$ cleaves the three boron–oxygen bonds in the addition product and forms $N(CH_2CH_2O)_3B$.

Figure B5.4

Figure B5.5

In each example described above the allyl group adds preferentially to one enantiotopic face of the aldehyde rather than the other. Interactions between the homochiral moiety attached to the allyl group and the aldehyde differ for each face of the aldehyde and the allyl group prefers to add to the face of the aldehyde for which these interactions (which may be due to a complex combination of steric and electronic factors) are minimized.

Synthesis of homoallylic alcohols containing two chiral centres

The efficient relay of stereochemical information in these reactions may be regarded as evidence supporting the involvement of cyclic transition states.

Addition of an aldehyde to an allylborane which is unsymmetrically substituted at the end of the carbon–carbon double bond remote from boron produces a homoallylic alcohol containing two adjacent chiral centres. (*E*)-Allylboranes give rise to the *threo* diastereoisomer whilst (*Z*)-allylboranes give rise to the *erythro* diastereoisomer. Figure B5.6 depicts the addition of two crotyldialkoxyboranes (derived from the appropriate stereoisomer of crotylmagnesium chloride and pinacol by a route analogous to the one used in Figure B5.4) to benzaldehyde.

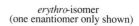

(*E*)-allylborane → 1. PhCHO / 2. N(CH₂CH₂OH)₃ →

threo-isomer
(one enantiomer only shown)

(*Z*)-allylborane → 1. PhCHO / 2. N(CH₂CH₂OH)₃ →

erythro-isomer
(one enantiomer only shown)

Figure B5.6

The stereochemical outcome of the reactions shown in Figure B5.6 is accounted for by the chair-like transition states shown in Figure B5.7. These place the aldehyde substituent in an equatorial position rather than an axial position, because in the latter position the substituent would give rise to unfavourable 1,3-diaxial interactions between itself and one of the oxygen substituents on the boron atom. Note that whether or not the methyl group occupies an equatorial or an axial position in the transition state is predefined by the double bond geometry of the allylborane.

The allylboranes used in Figure B5.6 are achiral and so only the relative stereochemistry of the products is controlled when they react with aldehydes. If homochiral allylboranes are used, however, then not only the relative

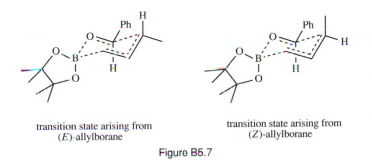

transition state arising from
(*E*)-allylborane

transition state arising from
(*Z*)-allylborane

Figure B5.7

stereochemistry but also the absolute stereochemistry of the products is controlled. This is illustrated by the generation of all four possible stereoisomers of 3-methyl-4-penten-2-ol from homochiral crotylboranes and ethanal (Figure B5.8). (The homochiral crotylboranes are synthesized from (+)- or (-)-α-pinene and (*E*)- or (*Z*)-crotylpotassium by routes analogous to the one used in Figure B5.3.)

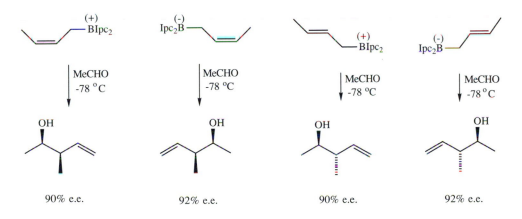

90% e.e.

92% e.e.

90% e.e.

92% e.e.

(+) and (-) refer to the optical rotation of the α-pinene used to generate the borane marked

Figure B5.8

B5.2 Boron enolates in organic synthesis

Boron enolates react with aldehydes and ketones under neutral conditions to give intermediates which hydrolyze to aldol products. The reaction proceeds *via* a cyclic transition state (Equation B5.2) and is analogous to the allylborane reactions discussed above.

The reaction proceeds both regioselectively and stereoselectively and has thus found many applications in organic synthesis.

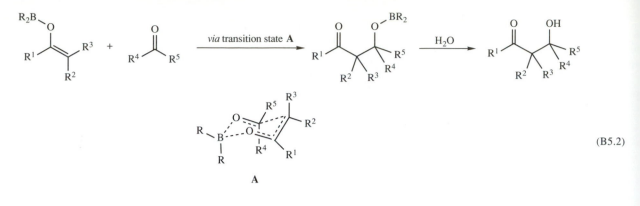

(B5.2)

A

Regioselectivity

By careful choice of base and dialkylboryl triflate it is possible to generate either the kinetic boron enolate or the thermodynamic boron enolate. These proceed to react with aldehydes without loss of regiochemical integrity, as shown in Equations B5.3 and B5.4.

$Bu_3B + HOTf \longrightarrow Bu_2BOTf$

$9\text{-BBN} + HOTf \longrightarrow 9\text{-BBNOTf}$

$Tf = \begin{matrix} O \\ \| \\ -S-CF_3 \\ \| \\ O \end{matrix}$

and ^-OTf is an excellent leaving group.

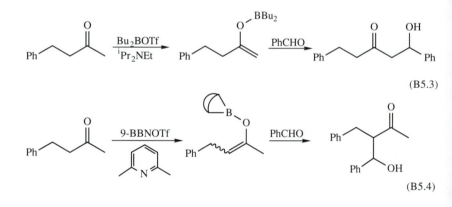

(B5.3)

(B5.4)

Diastereoselectivity

On addition of an aldehyde, (*E*)-enolates are converted to *threo* aldol products and (*Z*)-enolates are converted to *erythro* aldol products (Figure B5.9).

A chair-like transition state, in which the aldehyde substituent is placed in an equatorial position to prevent unfavourable 1,3-diaxial interactions with the axial boron substituent and the remote enolate substituent, explains these stereochemical results. The transition states for the reactions shown in Figure B5.9 are depicted in Figure B5.10.

Aldol reactions of boron enolates are frequently more diastereoselective than aldol reactions of, for example, lithium or aluminium enolates. This is partly ascribed to the relatively short boron–oxygen bond length (B–O = 1.36–1.47 Å, Li–O = 1.92–2.00 Å, Al–O = 1.92 Å) which exacerbates the unfavourable 1,3-diaxial interactions that occur between the boron substituent

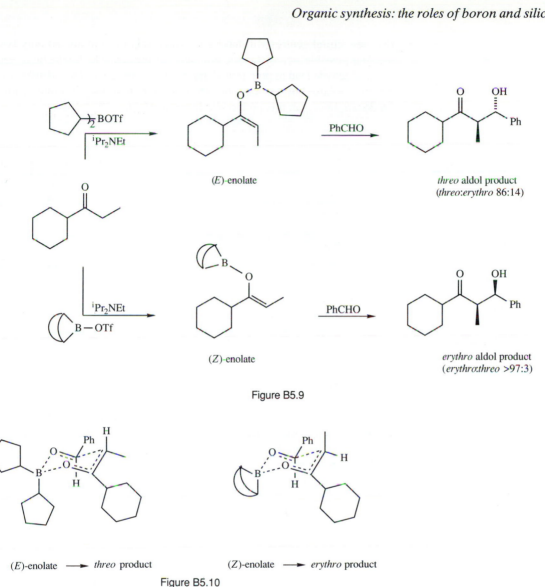

Figure B5.9

(E)-enolate ⟶ *threo* product   (Z)-enolate ⟶ *erythro* product

Figure B5.10

and the aldehyde substituent if the latter is in an axial position. (This increases the energy difference between the transition state in which the aldehyde substituent is equatorial, as shown in Figure B5.10, and the transition state in which the aldehyde substituent is axial; higher selectivity is therefore observed.)

Double asymmetric induction

Consider Equation B5.5 which depicts the reaction of a **homochiral aldehyde** with an **achiral boron enolate**. The aldol product contains two new chiral centres at C2 and C3. The boron enolate geometry dictates that the two new chiral centres will have an *erythro* relationship and so only two of the four possible stereoisomeric products are formed. The homochiral centres in the aldehyde lead to preferential approach to one face of the aldehyde by the boron enolate but, as can be seen by the resulting product ratio of $2R,3S:2S,3R = 3:2$, this preference is not marked.

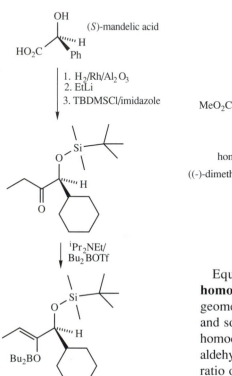

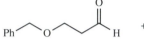

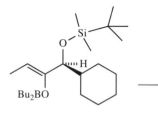

homochiral aldehyde   achiral boron enolate

((-)-dimethylglutaric hemialdehyde)

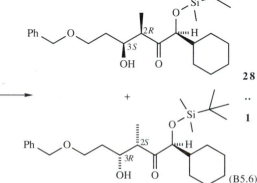

(B5.5)

Equation B5.6 depicts the reaction of an **achiral aldehyde** with a **homochiral boron enolate** derived from (*S*)-mandelic acid. Again the geometry of the enolate controls the relative stereochemistry of C2 and C3 and so only the two *erythro* isomers are formed. In this case, however, the homochiral centre in the boron enolate results in approach to one face of the aldehyde being strongly preferred over approach to the other face and a product ratio of $2R,3S:2S,3R = 28:1$ is observed.

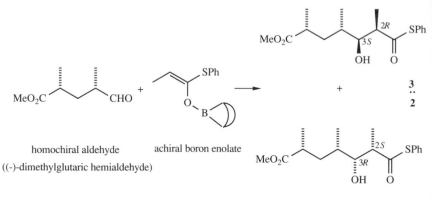

achiral aldehyde   homochiral boron enolate
derived from (*S*)-mandelic acid

(B5.6)

Thus the homochiral aldehyde (-)-dimethylglutaric hemialdehyde in Equation B5.5 and the homochiral (S)-boron enolate of Equation B5.6 both favour formation of the $2R,3S$ diastereoisomer. when these two molecules are reacted together their stereochemical preferences reinforce one another (they are said to be 'matched') and good stereoselectivity is observed ($2R,3S:2S,3R$ = >100:1) (Equation B5.7).

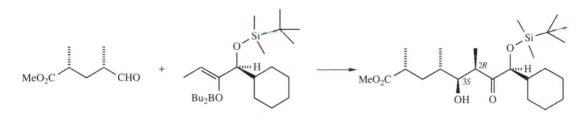

$$2R,3S : 2S,3R = >100:1$$

homochiral aldehyde   homochiral boron enolate
derived from (S)-mandelic acid

(B5.7)

If the boron enolate derived from (R)-mandelic acid is used in Equation B5.6 then a reversal of selectivity to $2R,3S:2S,3R$ = 1:28 is anticipated. Thus, on combination of the (R)-boron enolate with (-)-dimethylglutaric hemialdehyde, the stereochemical preferences of the two reagents work against each other (they are said to be 'mismatched') and relatively low stereoselectivity is observed. The stereochemistry of the major diastereoisomer is that produced by the partner with the strongest control over the reaction (Equation B5.8).

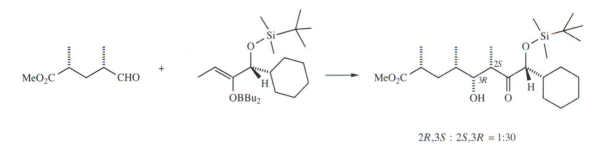

$$2R,3S : 2S,3R = 1:30$$

homochiral aldehyde   homochiral boron enolate
derived from (R)-mandelic acid

(B5.8)

Treatment of products derived from the mandelic acid-based boron enolates with aqueous HF followed by $NaIO_4$ releases a carboxylic acid. For example, the aldol product formed in Equation B5.7 is converted into a carboxylic acid (which lactonizes under the reaction conditions) as shown in Equation B5.9.

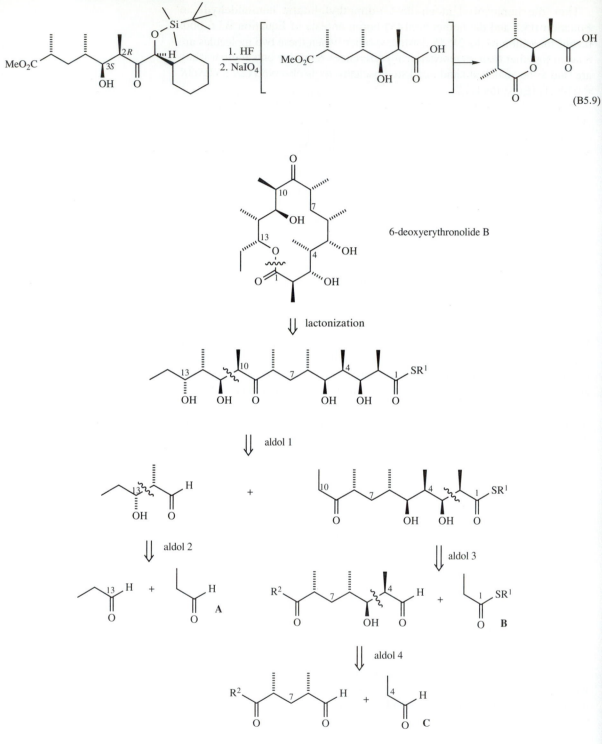

Figure B5.11

Thus the two mandelic acid-based boron enolates described in this section may be regarded as sources of propionic acid which add to aldehydes to give *erythro* aldol products of high stereochemical purity. An elegant synthesis of the macrolide, 6-deoxyerythronolide B, uses three mandelic acid-based boron enolate/aldehyde reactions. The retrosynthetic analysis of the synthesis is shown in Figure B5.11.

In the actual synthesis, aldol reaction 1 involved a lithium enolate whilst aldol reactions 2–4 used boron enolates based on the mandelic acid derivatives described above to incorporate fragments A, B, and C. The aldol products of reactions 2–4 were treated with fluoride and then $NaIO_4$ to release carboxylic acids which were derivatized or reduced as appropriate.

B6. Boronic ester homologation

Reaction of boronic esters, $RB(OR')_2$, with dichloromethyllithium, $LiCHCl_2$, inserts the $CHCl$ unit into the carbon–boron bond of the boronic ester. This is known as boronic ester homologation. If boronic esters derived from homochiral alcohols are used in this reaction, then new homochiral centres may be generated as will be illustrated below.

B6.1 Substrate synthesis

Expensive OsO_4 may be used in catalytic amounts to oxidize alkenes to diols if Me_3NO is added. This reagent efficiently reoxidizes the osmium back to OsO_4 during the course of the reaction.

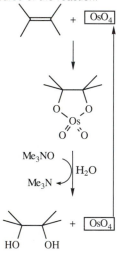

Many boronic ester homologation reactions have been performed using pinanediols as chiral auxiliaries. These are readily available from (+)- and (-)-α-pinene by osmium tetroxide-catalyzed oxidation reactions (Equations B6.1 and B6.2).

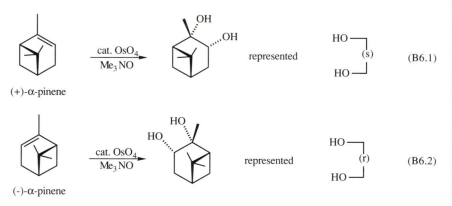

(+)-α-pinene

(B6.1)

(-)-α-pinene

(B6.2)

[By convention the (s) and (r) representations used in Equations B6.1 and B6.2 refer to the configuration of the α chiral centre in the α-chloroboronic esters that the homochiral pinanediol subsequently generates in the reactions described below.]

Addition of a wide range of boronic acids $[RB(OH)_2]$ or esters $[RB(OR')_2]$ to the pinanediols gives the very stable pinanediol boronic esters. For example, propylboronic acid (available from the addition of propylmagnesium bromide to trimethyl borate followed by acid hydrolysis) and the (s) pinanediol combine to give a homochiral boronic ester as shown in Equation B6.3.

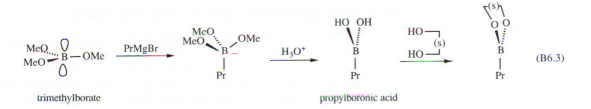

trimethylborate   propylboronic acid

B6.2 The dichloromethyllithium reaction

Figure B6.1 depicts the construction of a chiral centre on an (s) pinanediol boronic ester. Initially LiCHCl₂ is added to the substrate **A** to give a tetrahedral intermediate which on warming rearranges to the α-chloroboronic ester **B**. Under zinc chloride catalysis this rearrangement is highly stereoselective displacing only one of the chlorine atoms on the prochiral CHCl₂ group. Empirically the addition and rearrangement steps generate an *S* chiral centre when the chiral auxiliary is the pinanediol derived form (+)-α-pinene.

LiCHCl₂ is prepared from CH₂Cl₂ and ⁿBuLi.

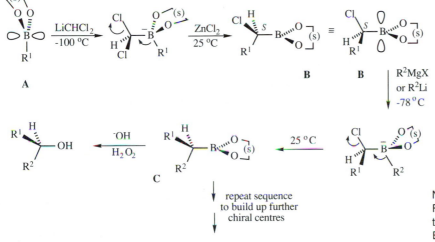

repeat sequence
to build up further
chiral centres

Note that the reactivity depicted in Figure B6.1 is a further example of the general reactivity described in Equation B2.2.

Figure B6.1

Subsequent addition of either an alkyllithium or a Grignard reagent to **B** leads to the displacement of the second chlorine atom with *clean inversion of configuration at the α centre*. The boronic ester may be oxidized at this point to generate a homochiral alcohol or, as **C** is merely an extension of **A**, the process may be repeated to build up further chiral centres as illustrated in the applications described below.

B6.3 Applications in asymmetric synthesis

Synthesis of (3*S*, 4*S*)-4-methylheptan-3-ol

(3*S*, 4*S*)-4-Methylheptan-3-ol is a component of the aggregation pheromone of the elm bark beetle *Scolytus multistriatus*. It is readily synthesized from the (s) pinanediol ester of propylboronic acid as shown in Figure B6.2.

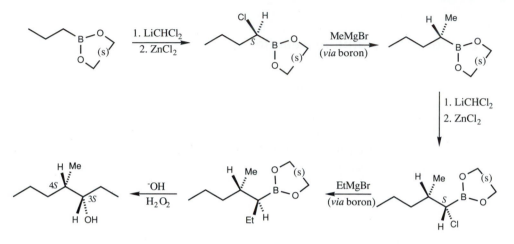

Figure B6.2

Synthesis of specifically deuterated phenylalanine

Boronic ester homologation has been used to synthesize specifically deuterated phenylalanine by the pathway shown in Figure B6.3.

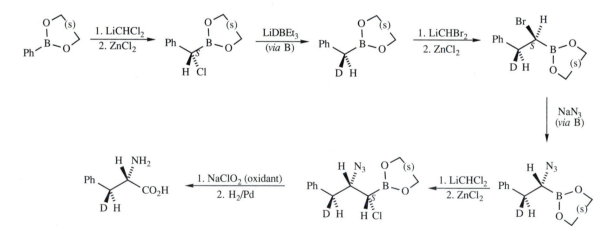

Figure B6.3

Synthesis of L-ribose

A boronic ester based synthesis of L-ribose, which contains three chiral
centres, is shown in Figure B6.4.

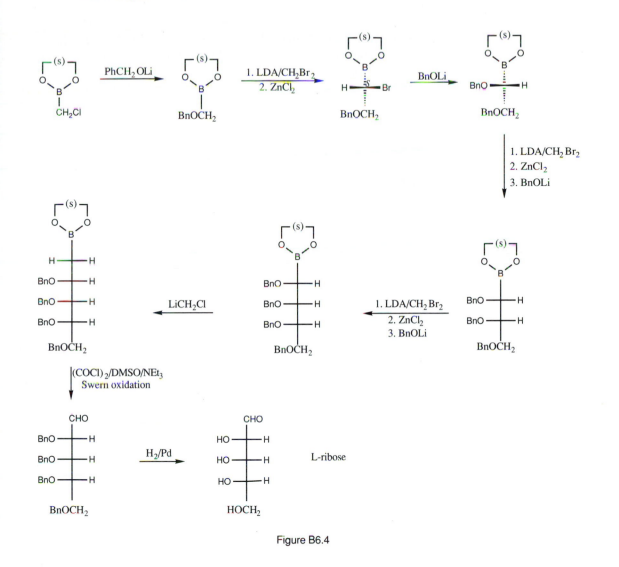

Figure B6.4

Further reading

Brown, H.C. and Jadhav, P.K. (1983). Asymmetric hydroboration. In *Asymmetric synthesis* (ed. J.D. Morrison), Vol. 2, pp 1–43. Academic Press, Orlando, Florida.

Brown, H.C., Negishi, E., and Zaidlewicz, M. (1982). Organoborane compounds in organic synthesis. In *Comprehensive organometallic chemistry* (ed. G. Wilkinson, F.G.A. Stone, and E.W. Abel), Vol.7, pp 111–363. Pergamon Press, Oxford.

Evans, D.A., Nelson, J.V., and Taber, T.R. (1982) Stereoselective aldol condensations. *Topics in Stereochemistry*, **13**, 1–115.

Hoffman, R.W. (1982). Diastereogenic addition of crotylmetal compounds to aldehydes. *Angewandte Chemie, International Edition in English*, **21**, 555–566.

Masamune, S., Choy, W., Peterson, J.S., and Sita, L.R. (1985). Double asymmetric synthesis and a new strategy for stereochemical control in organic synthesis. *Angewandte Chemie, International Edition in English*, **24**, 1–30.

Matteson, D.S. (1989). The use of chiral organoboranes in organic synthesis. *Synthesis*, 973–985.

Matteson, D.S. (1989). Boronic esters in stereodirected synthesis. Tetrahedron Report No. 250, *Tetrahedron*, **45**, 1859–1885.

Pelter, A., Smith, K., and Brown, H.C. (1988). *Borane reagents*. Academic Press, London.

Ramachandran, P.V. and Srebnik, M. (1987). The utility of chiral organoboranes in the preparation of optically active compounds. *Aldrichimica Acta*, **20**, 9–24.

Si1. Properties of organosilicon compounds

Si1.1 Properties of bonds to silicon

Bond strengths

The relative strengths of bonds that silicon and carbon form with some other elements are shown below. (The factors in parentheses indicate the approximate increase or decrease in strength between the silicon and carbon bond in question.)

Some typical bond dissociation energies (kJ mol^{-1}) for bonds to silicon and the corresponding bonds to carbon are given below.

Si–O	530	C–O	340
Si–F	810	C–F	450
Si–C	320	C–C	335

Si–O	>> C–O	(x 2.4–1.6)	Si–C	<	C–C	(x 0.95)
Si–F	>> C–F	(x 1.8)	Si–H	<	C–H	(x 0.85)
Si–Cl	> C–Cl	(x 1.4)				
Si–Br	> C–Br	(x 1.5)				
Si–I	> C–I	(x 1.5)				

It can be seen that silicon forms stronger bonds than carbon to oxygen and the halogens, and weaker bonds to carbon and hydrogen. The strength of silicon–oxygen and silicon–fluorine bonds cannot be over-emphasized: **much of organosilicon chemistry is driven by the formation of strong silicon–oxygen or silicon–fluorine bonds at the expense of other weaker bonds**. (It is of note that the silicon–fluorine bond is one of the strongest single bonds known.)

In contrast, π bonds between silicon and other elements are very weak and may be considered unimportant in applications of silicon to organic synthesis at present.

Note that the strength of the silicon–halogen bond decreases in the order:

Si–F > Si–Cl > Si–Br > Si–I

Bond lengths

The bonds between silicon and other atoms are generally significantly longer than those between carbon and the corresponding atoms. The relative increases in bond lengths between selected atoms attached to silicon and the corresponding bond to carbon are shown below.

Si–C	>	C–C	(x 1.25)
Si–H	>	C–H	(x 1.35)
Si–O	>	C–O	(x 1.15)

A typical Si–C bond length is 1.89 Å whilst a typical C–C bond length is 1.54Å.

There is some evidence that the trimethylsilyl group is less sterically demanding than its carbon analogue the *t*-butyl group. This is ascribed to the increased distance between the silicon atom and its point of attachment to the molecule in question.

Bond polarization

Silicon is considerably more electropositive than carbon and so carbon–silicon bonds are strongly polarized in the direction shown below.

$$\overset{\delta^+}{Si} - \overset{\delta^-}{C}$$

Si1.2 Nucleophilic substitution at silicon

Nucleophilic substitution at a silicon centre could in principle proceed by either an S_N2 or an S_N1 type mechanism. In practice the reaction normally follows an S_N2 type pathway sometimes referred to as the S_N2–Si pathway. The reaction follows this pathway not because of any inherent problem with an S_N1 type pathway, but because the S_N2–Si process is extremely rapid.

Almost all acyclic chloro-, bromo-, and iodosilanes react with almost all nucleophiles by the S_N2–Si mechanism which leads to inversion of configuration at silicon. For example, the homochiral chlorosilane depicted in Equation Si1.1 reacts with a range of nucleophiles including ethyllithium with clean inversion of configuration.

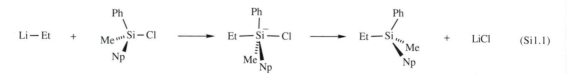

$$\text{(Si1.1)}$$

Np = 1-naphthyl

In many cases the pentaco-ordinate species involved in the reaction is thought to be an intermediate (i.e. the nucleophile and the leaving group are attached to the silicon by fully-formed bonds as depicted in Equation Si1.1) rather than a transition state as invoked in the analogous carbon S_N2 reaction.

Si1.3 Stabilization of β-carbocations and α-carbon–metal bonds

Many reactions encountered in organosilicon chemistry involve the formation of carbocations on a carbon atom β to a silicon atom (Si–C–C$^+$) and carbanions on a carbon atom α to a silicon atom (Si–C$^-$).

Stabilization of β-carbocations

The so-called β-effect has been ascribed to overlap between the vacant p orbital on the β carbon atom and the σ orbital between the silicon atom and the α carbon atom (Figure Si1.1). (Due to the electronegativities of the atoms involved, the carbon–silicon σ-orbital has a higher coefficient on carbon.)

Maximum stabilization only occurs of course if the vacant p orbital and the carbon–silicon bond are in the same plane. Whilst this does not present any problems in acyclic cases, it is not always possible in cyclic systems.

Stabilization of α-carbon–metal bonds

The ability of silicon to stabilize an α-carbanion has been attributed to several factors including (a) overlap of the α-carbon–metal bond with a silicon d orbital, and (b) overlap of the α-carbon–metal bond with the σ* orbital of an adjacent silicon–carbon bond (Figure Si1.2). (The carbon–silicon σ* orbital has a higher coefficient on silicon.)

Figure Si1.1

The increased stability of a carbocation β to silicon determines the regioselectivity of the reactions between electrophiles and both allylsilanes and vinylsilanes (see p. 84 and p. 76 respectively).

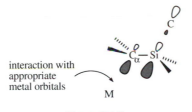

interaction with appropriate metal orbitals

M

Figure Si1.2

The stability of carbanions α to silicon facilitates nucleophilic attack on vinylsilanes (see p. 78).

Si1.4 1,2-Rearrangements

When silylcarbinols are treated with catalytic amounts of base or active metal, the silyl group migrates from carbon to oxygen (Equations Si1.2 and Si1.3). Detailed studies of this reaction have been carried out by A.G. Brook whose name is now associated with the reaction.

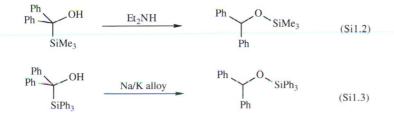

The reaction pathway is outlined in Figure Si1.3.

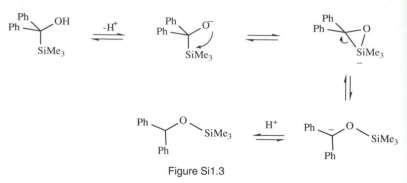

Figure Si1.3

The reverse of the Brook rearrangement is sometimes referred to as the silyl-Wittig rearrangement. Although the overall transformation is analogous to the Wittig rearrangement shown below, it should be noted that the Wittig rearrangement proceeds *via* a radical anion cage mechanism.

The reaction is an equilibrium with the strength of the silicon–oxygen bond normally favouring the silyl ether. In some cases, however, the reverse reaction may be observed. An example of such a reaction is given in Equation Si1.4.

(Si1.4)

Si2. Protection of hydroxy groups as silyl ethers

Si2.1 Formation of silyl ethers

The trimethylsilyl group has been used extensively for the protection of alcohols. One of the many methods which have been used for protecting a hydroxy group as its trimethylsilyl ether involves adding trimethylsilyl chloride (trimethylchlorosilane, TMCS) to the alcohol in the presence of a weak base as exemplified in Equation Si2.1.

$$(Si\,2.1)$$

Many other reagents have been developed for converting hydroxy groups to trimethylsilyl ethers more efficiently or more selectively; two of these reagents are depicted in Figure Si2.1.

TMSDEA	TMSI
(*N*-trimethylsilyldiethylamine)	(*N*-trimethylsilylimidazole)

The order of reactivity of these reagents towards simple hydroxy groups is:

TMSI > TMSDEA > TMCS/pyridine

Figure Si2.1

The level of selectivity that can be achieved in the formation of trimethylsilyl ethers is illustrated by the selective protection of the hydroxy group at C-11 in the methyl ester of the prostaglandin 15-methyl PGF$_{2\alpha}$ (Equation Si2.2). Thus the less sterically hindered secondary alcohol at C-11 is selectively protected in the presence of the more sterically hindered secondary alcohol at C-9 and the tertiary alcohol at C-15.

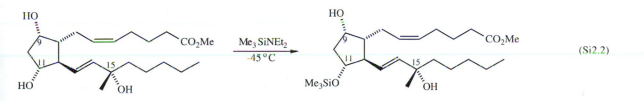

$$(Si2.2)$$

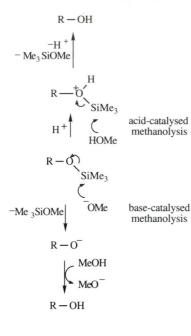

Trimethylsilyl ethers are sensitive to certain reagents used in organic synthesis. For example, they are attacked by nucleophiles, readily cleaved under acidic or basic conditions, and they do not survive hydrogenolysis. To overcome these problems a number of reagents which form more bulky silyl ethers have been developed; some of these are shown in Figure Si2.2.

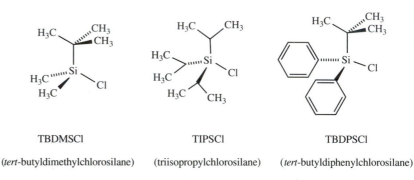

TBDMSCl   TIPSCl   TBDPSCl

(*tert*-butyldimethylchlorosilane)   (triisopropylchlorosilane)   (*tert*-butyldiphenylchlorosilane)

Figure Si2.2

Tert-butyldimethylsilyl ethers have been used extensively for the protection of hydroxy groups. They are more stable to hydrolysis than trimethylsilyl ethers by a rate factor of 10^4 and they are compatible with a much wider range of reagents used in organic synthesis.

The steric bulk which accounts for the relative stability of *tert*-butyldimethylsilyl ethers also hinders their formation and so the reaction between *tert*-butyldimethylsilyl chloride and hydroxy groups in the presence of pyridine is very slow. In the presence of catalytic amounts of imidazole, however, the reaction proceeds rapidly and in high yield as shown in Equation Si2.3.

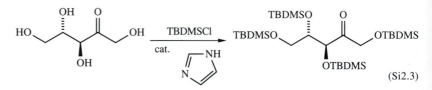

(Si2.3)

The TBDMSCl reacts with imidazole to give the intermediate depicted in Figure Si2.3. In its protonated form the intermediate is a highly reactive silylating reagent.

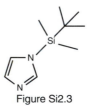

Figure Si2.3

Si2.2 Cleavage of silyl ethers

Silyl ethers are cleaved to their parent alcohols by nucleophiles (often alcohols) under a range of acidic or basic conditions (Equation Si2.4).

(Si2.4)

The rate of cleavage is inversely proportional to the bulk of the silyl ether and so trimethylsilyl groups are removed relatively easily whilst *tert*-butyldimethylsilyl groups are more resistant to removal.

The fluoride ion is used extensively for cleaving silyl ethers (Equation Si2.5). Reagents used include Bu₄NF (tetra-*n*-butylammonium fluoride, TBAF) and KF.18-crown-6.

MeO⁻/MeOH cleavage of tri-methylsilyl ethers occurs much more rapidly (by a factor of approximately 10^4) than the corresponding cleavage of *tert*-butyldimethylsilyl ethers. Both types of ether, however, are very rapidly cleaved by F⁻.

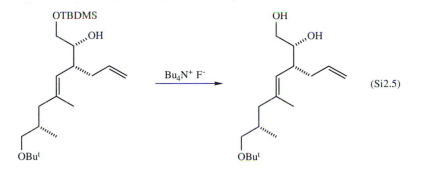

(Si2.5)

Si2.3 Examples of protection of hydroxy groups as silyl ethers

(a) Figure Si2.4 outlines a synthesis of (*R*)-isoproterenol, a compound which displays β-adrenoreceptor activity. Note that this synthesis uses two types of silicon-based protecting groups.

(b) During a synthesis of gephyrotoxin, one of a variety of alkaloids isolated from the skin of the Colombian frog *Dendrobates histrionicus*, it was found that the bulky *tert*-butyldiphenylsilyl group could be used to direct hydrogen delivery to the opposite face of a tricyclic skeleton and thus control the stereochemical outcome of a hydrogenation step (Figure Si2.5).

By co-ordinating BH₃ to its nitrogen atom and the oxygen atom of the ketone to be reduced to its boron atom, the homochiral catalyst **A**, depicted below, facilitates the delivery of hydride to one face of the prochiral ketone.

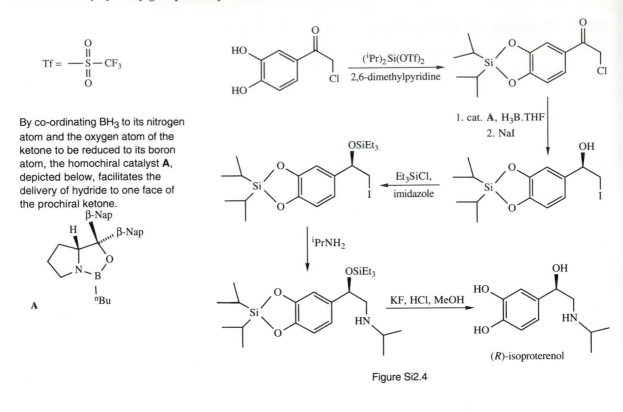

Figure Si2.4

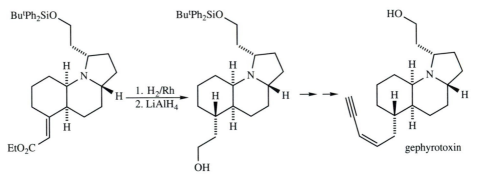

Figure Si2.5

Si3. Silyl enol ethers and related silyl ethers

Si3.1 Silyl enol ethers

Silyl enol ethers (Figure Si3.1) are stable molecules which may be isolated, purified, and characterized using standard organic procedures. As will be illustrated below, **they are sources of regiochemically-pure enolate ions and their equivalents**, and as such they play an important role in modern synthetic organic chemistry.

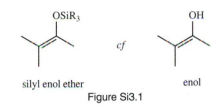

silyl enol ether   *cf*   enol

Figure Si3.1

Formation of silyl enol ethers

(i) *Trapping of enolate anions*
Silyl enol ethers are generally prepared from enolate ions as illustrated in Equation Si3.1.

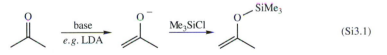

(Si3.1)

Deprotonation of an *unsymmetrically substituted ketone* such as 2-methylcyclohexanone (Figure Si3.2) potentially gives rise to two isomeric enolate anions. Under kinetic conditions, deprotonation at the least substituted carbon atom is favoured and the enolate anion with the least substituted double bond is in excess. Under thermodynamic conditions however, equilibration between the two enolate anions occurs and the enolate anion with the more substituted double bond eventually predominates.

Addition of a silylating reagent such as Me_3SiCl to the reaction mixture traps the enolate anions and produces two silyl enol ethers in a ratio which reflects the ratio of the enolate anions. Thus if 2-methylcyclohexanone is added to the hindered base LDA at -78 °C and the mixture stirred for 1 hour at -78 °C and quenched with Me_3SiCl, then the major product is the silyl enol ether derived from the 'kinetic enolate'. In contrast, heating 2-methylcyclohexanone, triethylamine, and Me_3SiCl at 130 °C for 90 hours

> Recall that the stability of double bonds increases with increasing substitution.

gives the silyl enol ether of the 'thermodynamic enolate' as the major product (Figure Si3.2).

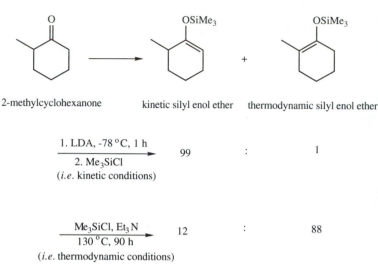

2-methylcyclohexanone   kinetic silyl enol ether   thermodynamic silyl enol ether

The kinetic and thermodynamic silyl enol ethers generated in reactions analogous to those depicted in Figure Si3.2 may be separated by distillation, crystallization, etc. Thus pure samples of each silyl enol ether may be obtained and these may be used to generate pure enolates as described below.

1. LDA, -78 °C, 1 h
———————————————→   99   :   1
2. Me₃SiCl
(*i.e.* kinetic conditions)

Me₃SiCl, Et₃N
———————————————→   12   :   88
130 °C, 90 h
(*i.e.* thermodynamic conditions)

Figure Si3.2

(ii) *Conjugate addition/silylation*

Conjugate addition to an α,β-unsaturated ketone followed by silylation may be used to generate a regiochemically-pure silyl enol ether (Equation Si3.2).

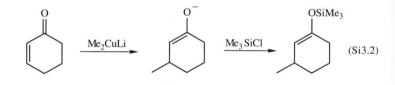

(Si3.2)

(iii) *Hydrosilylation*

Under rhodium catalysis, hydridosilanes add to α,β-unsaturated ketones in a 1,4 manner to produce silyl enol ethers (Equation Si3.3).

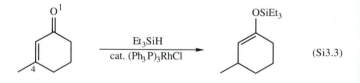

(Si3.3)

Reactions of silyl enol ethers

Addition of methyllithium to a trimethylsilyl enol ether gives the corresponding lithium enolate and volatile tetramethylsilane (Equation Si3.4).

(Si3.4)

Similarly, benzyltrimethylammonium fluoride cleaves trimethylsilyl enol ethers to give quaternary ammonium enolates and Me₃SiF (Equation Si3.5).

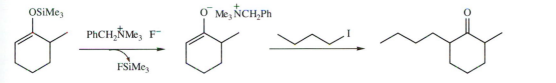

(Si3.5)

The enolate anions are more reactive towards electrophiles when they are associated with non-coordinating quaternary ammonium cations than when they are associated with lithium cations. Thus, as illustrated in Equations Si3.4 and Si3.5, quaternary ammonium derivatives are preferred as counterions for kinetic enolates in order to prevent any isomerization to the thermodynamic enolate occurring before reaction with the added electrophile proceeds.

A second important reaction of silyl enol ethers is their reaction with strong electrophiles, which is depicted schematically in Figure Si3.3.

Enolates, enamines, and silyl enol ethers all exhibit similar reactivity towards electrophiles.

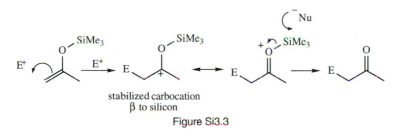

stabilized carbocation
β to silicon

Figure Si3.3

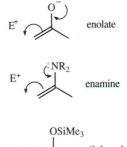

Silyl enol ethers react cleanly with a wide range of electrophiles, including tertiary halides, if the electrophilicity of the alkylating reagent is enhanced by the presence of a Lewis acid. Generation of two contiguous quaternary centres is even possible as shown in Figure Si3.4. (Alkylation of enolates with tertiary halides is normally prevented by competing elimination reactions.)

Amongst the electrophiles which may be added to trimethylsilyl enol ethers under Lewis acid catalysis are aldehydes and ketones (Equation Si3.6–8).

Silyl enol ethers alone, however, are non-basic and react exclusively on carbon.

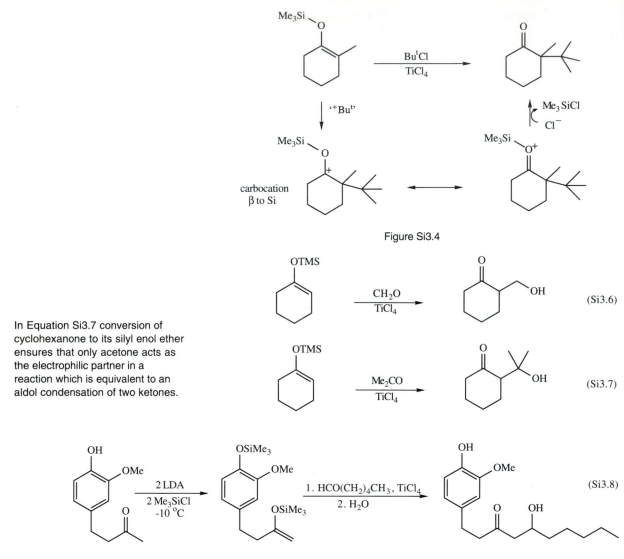

Figure Si3.4

In Equation Si3.7 conversion of cyclohexanone to its silyl enol ether ensures that only acetone acts as the electrophilic partner in a reaction which is equivalent to an aldol condensation of two ketones.

(Si3.6)

(Si3.7)

(Si3.8)

gingerol, a constituent of ginger

Si3.2   2-Trimethylsilyloxybuta-1,3-dienes

2-Trimethylsilyloxybuta-1,3-dienes (Figure Si3.5) may be used as the 4π component in Diels–Alder reactions and in this capacity they have been incorporated into many major syntheses.

Me$_3$SiO

Figure Si3.5

They are generally prepared by silylation of enolate anions derived from α,β-unsaturated carbonyl compounds (Equations Si3.9 and 3.10).

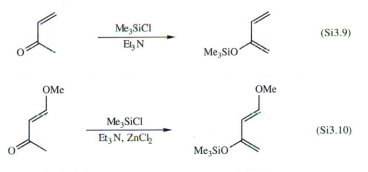

(Si3.9)

(Si3.10)

Danishefsky's diene

It is of note that due to their inherent polarity, 2-trimethylsilyloxybuta-1,3-dienes undergo regiospecific Diels–Alder reactions. As exemplified in Equation Si3.11, hydrolysis of the trimethylsilyl enol ether present in the Diels–Alder product reveals a substituted cyclohexanone.

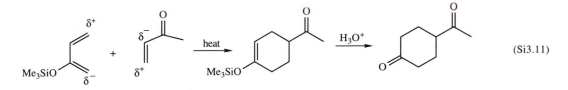

(Si3.11)

Like other Diels–Alder reactions, reactions involving 2-trimethylsilyloxy-buta-1,3-dienes proceed under good stereochemical control. In the examples depicted in Equation Si3.12 and Si3.13, the dienophiles react with the dienes through conventional *endo* transition states to give exclusively *endo* adducts.

Recall that the *endo* transition state of the Diels–Alder reaction is stabilized by favourable secondary orbital interactions between the π systems of the diene and the dienophile.

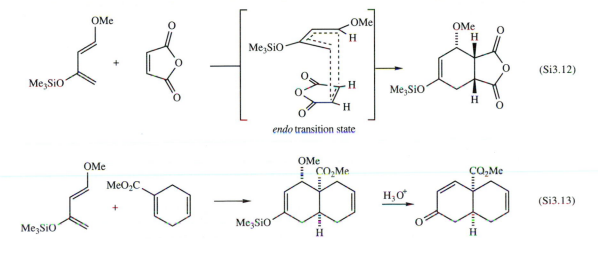

endo transition state

(Si3.12)

(Si3.13)

Under Lewis acid catalysis, 1-methoxy-3-trimethylsilyloxybuta-1,3-dienes react with aldehydes to give dihydropyrones (Equation Si3.14).

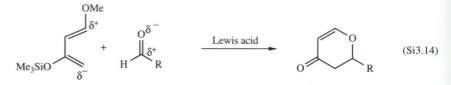

(Si3.14)

Two extreme pathways may be envisaged for this transformation. One involves a pericyclic reaction followed by desilylation and elimination of methoxide from the product silyl enol ether. The alternative proceeds *via* a Lewis acid promoted aldol reaction to give an intermediate which cyclizes to the product dihydropyrone (Figure Si3.6). The actual pathway in an individual case is affected to a considerable degree by the nature of the Lewis acid used and lies somewhere between these two extremes.

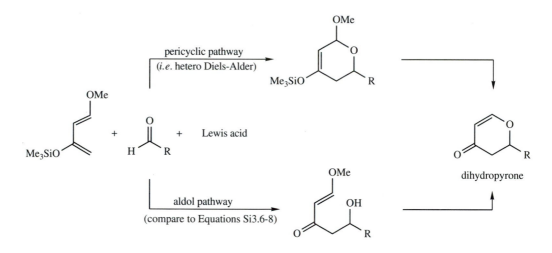

Figure Si3.6

cis co-ordination is less stable than *trans* co-ordination

Addition of a substituent onto the diene terminus remote from the methoxy group ultimately introduces a second chiral centre into the product dihyropyrone. Under zinc chloride catalysis the relative stereochemistry of these two centres is observed to be *cis* which is consistent with an *endo* transition state and the pericyclic pathway (Figure Si3.7). *Cis* products are obtained for a range of aldehydes including simple aliphatic aldehydes and so secondary orbital interactions do not adequately explain all the results obtained. It is postulated therefore that the bulky catalyst/solvent ensemble co-ordinates to the aldehyde *trans* to its substituent. This places the substituent in the *endo* position which leads to *cis* stereochemistry in the product.

Introduction of an alkoxy group α to the aldehyde results in a dramatic change from *endo* to *exo* topology (Figure Si3.8). Here chelation between the bulky catalyst and the aldehyde substituent places them in a *cis* relationship

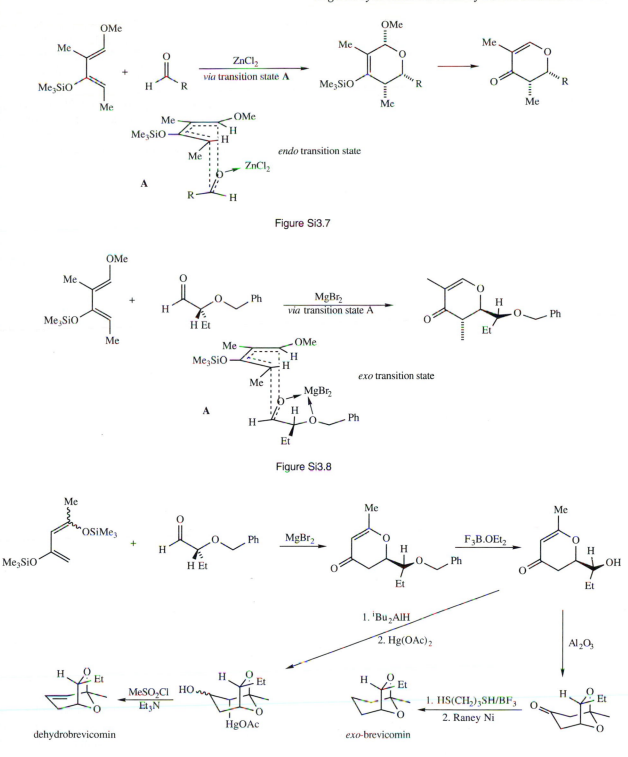

Figure Si3.7

Figure Si3.8

dehydrobrevicomin

exo-brevicomin

Figure Si3.9

and so the aldehyde lies in the *exo* orientation. Furthermore, in the case illustrated in Figure Si3.8 the chiral carbon atom in the aldehyde substituent renders the faces of the aldehyde diastereotopic. The sterically more demanding ethyl group is thus placed away from the incoming diene in the transition state.

If the reaction depicted in Figure Si3.8 is performed with a slightly modified diene, a dihydropyran is formed which provides the framework for *exo*-brevicomin, a pheromone of the western pine beetle *Dendroctonus brevicomis*, and dehydrobrevicomin, a compound which promotes agression in male mice (Figure Si3.9).

Si3.3   Silyl ketene acetals

Figure Si3.10

Silyl ketene acetals (Figure Si3.10) are derived from ester enolates and are closely related to silyl enol ethers (Figure Si3.1).

They may be synthezised by deprotonation and silylation of carboxylic esters or lactones (Equations Si3.15 and Si3.16). Note that in the example depicted in Equation Si3.15 the *E* isomer is generated if the reaction is performed in THF, but addition of polar HMPA [hexamethylphosphoric triamide, $(Me_2N)_3P{=}O$] to the reaction mixture gives the *Z* isomer.

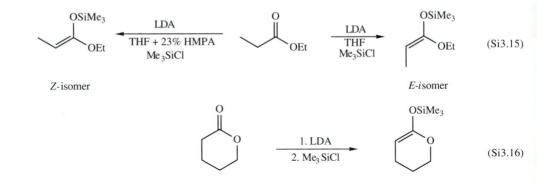

Z-isomer E-*isomer*

(Si3.15)

(Si3.16)

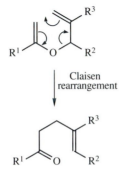

Claisen
rearrangement

Silyl ketene acetals derived from allyl esters undergo [3,3]-sigmatropic Claisen rearrangements and this reaction has been applied to a range of synthetic problems. Incorporation of an electron-donating group into the vinyl group of an allyl vinyl ether increases the rate at which the allyl vinyl ether undergoes the Claisen rearrangement. Thus the silyl ketene acetals derived from allyl esters rearrange readily at room temperature or just above to give γ,δ-unsaturated acids after desilylation (Equation Si3.17).

During the rearrangement a new carbon–carbon double bond and a new carbon–carbon single bond are formed. With appropriate substituents, the stereochemistry about these bonds becomes important and may be predicted accurately by assuming that the rearrangement passes through a chair-like transition state. For example, a substituent adjacent to the oxygen atom on the allyl side of the ether leads to an *E* double bond in the product (Equation Si3.18).

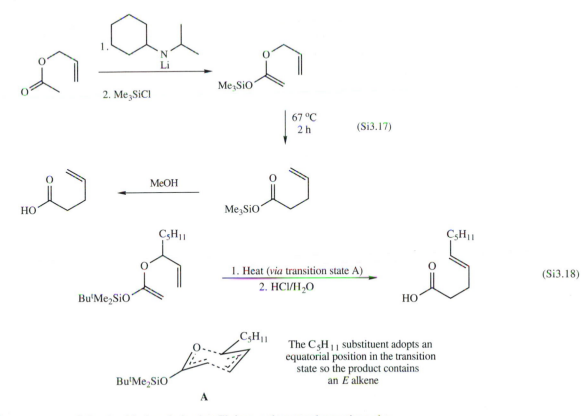

(Si3.17)

(Si3.18)

The C$_5$H$_{11}$ substituent adopts an equatorial position in the transition state so the product contains an *E* alkene

The geometry of the double bonds in the Claisen substrate determines the stereochemistry around the newly-formed carbon–carbon single bond in the product. For example (*E*)- and (*Z*)-silyl ketene acetals produce diastereoisomeric products as illustrated in Figure Si3.11.

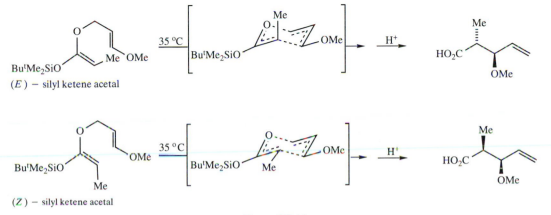

Figure Si3.11

Claisen rearrangements of silyl ketene acetals have been used in numerous syntheses, including the synthesis of the ionophore antibiotic Calcimycin outlined in Figure Si3.12.

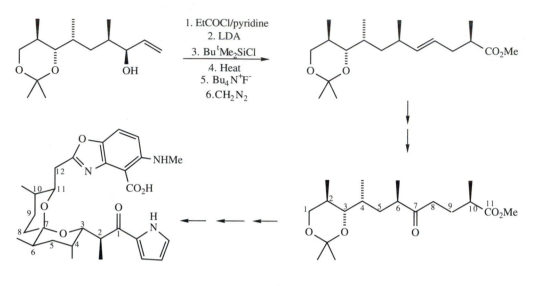

Figure Si3.12

Si3.4 1,2-bis-siloxy alkenes

Carboxylic esters react with sodium metal to give α-hydroxy ketones (often referred to as acyloins). The reaction, known as the acyloin condensation, is thought to proceed by the mechanism shown in Figure Si3.13.

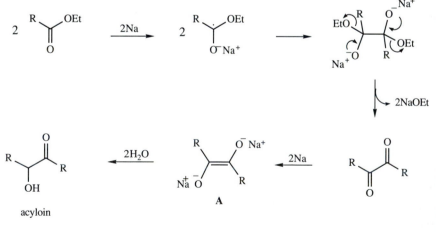

Figure Si3.13

A serious problem in the acyloin condensation is the generation of NaOEt. The ethoxide produced catalyzes several side-reactions including carbonyl condensations such as the Dieckmann condensation. Addition of more than four equivalents of trimethylsilyl chloride to the reaction (the Rühlmann modification) serves two purposes: (*i*) it reacts with any ethoxide generated to give the ether Me_3SiOEt and so minimizes side-reactions; (*ii*) it traps out the unsaturated dianion intermediate in the reaction (**A** in Figure Si3.13) to form 1,2-bis-siloxy alkenes, which not only undergo acid-catalysed hydrolysis or alcoholysis to the appropriate acyloin but also partake in other interesting reactions. Thus if sodium is added to diethyl adipate the reaction follows a Dieckmann condensation pathway, whereas if trimethylsilyl chloride is added to the reaction mixture the Dieckmann condensation is inhibited and an acyloin is produced (Figure Si3.14).

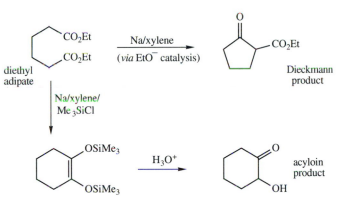

In the absence of Me_3SiCl, the Dieckmann condensation in Figure Si3.14 proceeds faster than the corresponding acyloin condensation.

Figure Si3.14

The silicon-modified acyloin condensation has been used widely to construct four-membered rings and their derivatives from 1,4-diesters (Equation Si3.19–20).

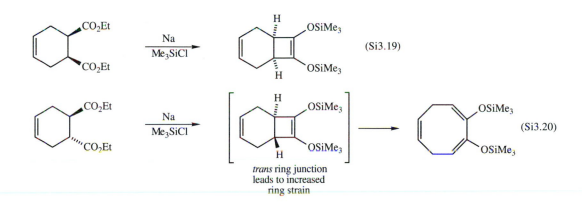

Products from silicon-modified acyloin condensations have also been used successfully in syntheses of macrocylic lactones. For example, decan-9-olide (a defensive secretion of *Phoracantha synonyma*) has been synthezised from diethyl adipate by the route shown in Figure Si3.15. The 1,2-bis-siloxy alkene product was treated with two equivalents of methyllithium to give a dianion to which 3-hydroxy-1-iodobutane was added. Cleavage of the resultant diol using Pb(OAc)$_4$ followed by ketone reduction completed the synthesis of decan-9-olide.

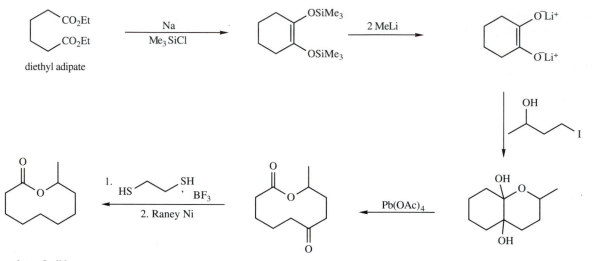

Figure Si3.15

Si4. Alkene synthesis (Peterson olefination)

Si4.1 Routes to β-hydroxysilanes

β-Hydroxysilanes are key compounds in the application of organosilicon chemistry to the synthesis of alkenes. Three methods of generating them are described below.

1. Addition of α-metallated silanes to carbonyl compounds

α-Metallated silanes may be generated in several ways including (a) addition of organometallic nucleophiles to vinylsilanes, (b) deprotonation α to silicon, and (c) metal–halogen exchange. Equations Si4.1–3 depict the generation of α-metallated silanes by each of these methods and their subsequent addition to carbonyl compounds to generate β-hydroxysilanes.

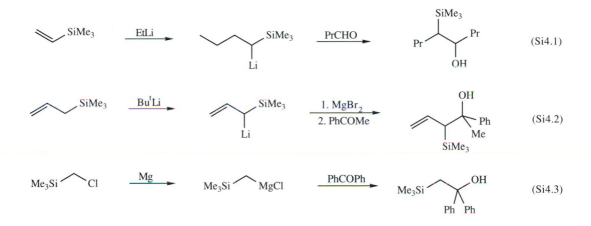

$$(Si4.1)$$

$$(Si4.2)$$

$$(Si4.3)$$

2. Addition of dialkylcuprates to α,β-epoxysilanes

α,β-Epoxysilanes are readily synthesized by epoxidation of vinyl silanes (see, e.g., Equations Si5.15 and 5.16). They react with organocuprate reagents in a regiospecific manner to give β-hydroxysilanes as exemplified in Equation Si4.4.

$$(Si4.4)$$

Note that the epoxide in Equation Si4.4 is opened at its most hindered end as the silyl group promotes substitution α to itself (see also Equations Si5.18 and 5.19).

3. Addition of hydride sources or organometallic reagents to β-ketosilanes

β-Ketosilanes react with hydride sources or organometallic reagents to give β-hydroxysilanes (Equation Si4.5). (Trimethylsilylacetone is prepared by reaction of Me_3SiCH_2MgCl with acetic anhydride, whilst more substituted β-ketosilanes may be formed by oxidation of β-hydroxysilanes.)

(Si4.5)

Si4.2 Conversion of β-hydroxysilanes into alkenes

Elimination of the trialkylsilyl group and the hydroxy group from β-hydroxysilanes leads to alkenes. **The reaction follows two distinct mechanistic pathways depending on whether the elimination is carried out under acidic or basic conditions.** Under acidic conditions *anti*-elimination occurs, whilst under basic conditions *syn*-elimination occurs. Thus the *threo* diastereoisomer of 5-trimethylsilyloctan-4-ol gives *cis* oct-4-ene when treated with H_2SO_4 or BF_3 and *trans* oct-4-ene on stirring with NaH or KH (Figure Si4.1).

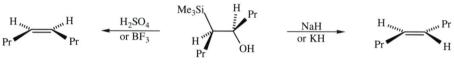

threo-5-trimethylsilyloctan-4-ol

Figure Si4.1

Elimination under acidic conditions requires the trimethylsilyl and hydroxy groups to be in an antiperiplanar relationship as shown in Figure Si4.2. This requirement is consistent with the observation that the cyclic β-hydroxysilane, also depicted in Figure Si4.2, is stable under acidic conditions.

The antiperiplanar relationship of the trimethylsilyl group and the hydroxy group places the bonds to be broken or formed in the same plane and thus maximizes orbital overlap.

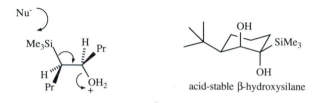

acid-stable β-hydroxysilane

Figure Si4.2

Elimination under basic conditions requires that the trimethylsilyl and hydroxy groups adopt a synperiplanar relationship as shown in Figure Si4.3.

This facilitates the formation of a strong silicon–oxygen bond and an intermediate four-membered ring which breaks down in a manner analogous to the final step in a Wittig reaction. It is of note that overall a strong silicon–oxygen bond replaces a weaker silicon–carbon bond which overrides the replacement of a strong carbon–oxygen bond with a weaker carbon–carbon π bond.

Figure Si4.3

The *threo* diastereoisomer of 5-trimethylsilyloctan-4-ol, used in Figures Si4.1–3 to illustrate alkene synthesis from β-hydroxysilanes, may be synthesized by the two routes illustrated in Figure Si4.4. In the first approach, a regiospecific and stereospecific opening of an epoxysilane (one enantiomer of racemate used shown for clarity) by an organocuprate reagent is used to generate the β-hydroxysilane. In the second approach, addition of ethyllithium to trimethylvinylsilane is used to generate an α-silylated carbanion which is quenched with carbon dioxide to form a carboxylic acid (one enantiomer shown for clarity). The carboxylic acid is converted into the propyl ketone depicted *via* its acid chloride. Finally, reduction of the ketone using DIBAL (^{i}Bu^{2}AlH) occurs in the manner predicted by the Felkin-Anh model to give *threo* 5-trimethylsilyloctan-4-ol.

Recall that the Felkin-Anh model predicts the preferred approach of a nucleophile to a carbonyl group bearing a chiral centre. The model places the most bulky group on the chiral centre (L) perpendicular to the plane of the carbonyl and its substituent (R), and the least bulky group on the chiral centre (S) adjacent to the carbonyl substituent (R). The nucleophile then prefers to approach the carbonyl past the smallest substituent on the chiral centre.

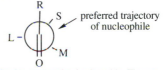

In the example depicted in Figure Si4.4, R = Pr, L = Me$_3$Si, M = Pr, and S = H.

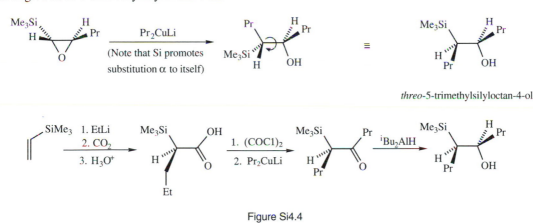

Figure Si4.4

Si4.3 Examples of silicon-based alkene synthesis

Silicon-based alkene synthesis is used much less than the analogous phosphorus-based Wittig reaction. In some cases, however, the silicon-based method has proved superior to the phosphorus-based chemistry. For example, a Wittig reaction failed to methylenate the hindered ketone shown in Equation

Si4.6 whereas a silicon-based approach proved to be successful and gave the required exocyclic double bond in β-Gorgonene, a non-isoprenoid sesquiterpene isolated from *Pseudopteragorgia americana*. The success of the α-silyl carbanion here is attributed to the lengthy carbon–silicon bond which makes the carbanion sterically less demanding.

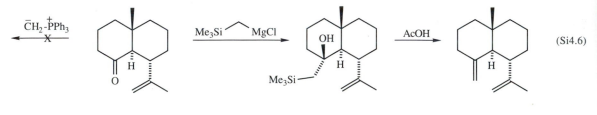

(Si4.6)

β-Gorgonene

Similarly in a synthesis of warburganal, a sesquiterpene isolated from certain tree barks which possesses strong antifeedant activity as well as molluscicidal and antimicrobial properties, conversion of a highly hindered ketone into an alkene failed with several phosphorus-based reagents. Addition of lithium methoxy(trimethylsilyl)methylide, however, proceeded satisfactorily, although without control over the formation of the chiral centre adjacent to the silyl group and hence without control over the geometry of the double bond subsequently produced (Figure Si4.5).

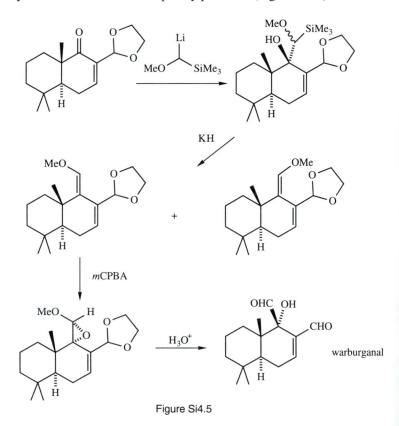

Figure Si4.5

Si5. Alkynyl-, vinyl-, and arylsilanes

Si5.1   Alkynylsilanes

Terminal alkynes are converted into alkynylsilanes simply by metallation and addition of a chlorosilane. The reverse transformation may be effected by several reagents including NaOH, $Bu_4N^+F^-$, and $AgNO_3$ followed by KCN (Figure Si5.1).

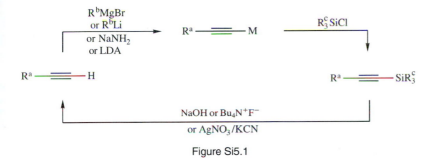

Figure Si5.1

Alkynylsilanes as protected alkynes

Conversion of a terminal alkyne to its alkynylsilane prevents loss of the relatively acidic terminal hydrogen (pK_a of ethyne c. 25) during later synthetic steps. For example, the terminal hydrogen of propyne was masked whilst its propargylic anion was used in a synthesis of Cecropia juvenile hormone, a chemical which plays an important role in insect development (Figure Si5.2).

Reactions of internal alkynes may be performed in the presence of terminal alkynes if the terminal alkyne is first converted to an alkynylsilane. In the example depicted in Figure Si5.3, hydrogenation of a silylated diyne using the Lindlar catalyst partially reduces the non-silylated alkyne and leaves the sterically hindered alkynylsilane intact.

Regioselective electrophilic attack on alkynylsilanes

Electrophiles react with alkynylsilanes at the silicon-bearing end of the alkyne thus generating positive character β to the silicon atom (Figure Si5.4).

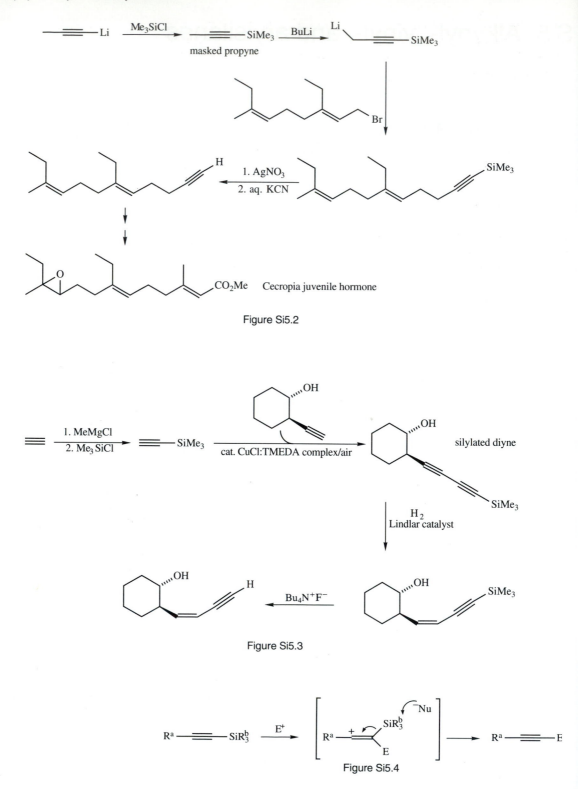

Figure Si5.2

Figure Si5.3

Figure Si5.4

Thus alkynylsilanes combine with acid chlorides to give alkynyl ketones in an electrophilic substitution reaction which is exemplified in Equation Si5.1 by the Lewis acid-catalysed reaction between 1-trimethylsilylhex-1-yne and ethanoyl chloride.

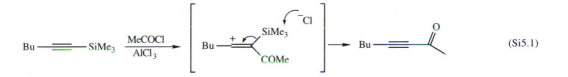

(Si5.1)

A dramatic illustration of the effect of a trialkylsilyl group on the reactivity of alkynes is provided by the cyclization reactions of the trienynes depicted in Figure Si5.5. When a methyl group terminates the trienyne system, the cyclization proceeds to give a steroid containing a five-membered D-ring. In contrast, a terminal trimethylsilyl group on the trienyne alters the course of the acid-promoted cyclization, and a steroid containing a six-membered D-ring is produced.

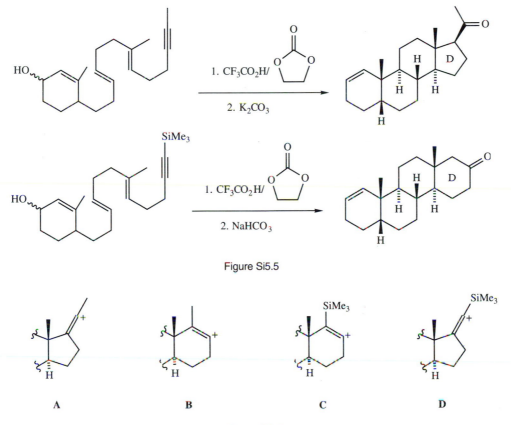

Figure Si5.5

A   B   C   D

Figure Si5.6

It is proposed that cyclization of the methyl trienyne proceeds *via* formation of the preferred linear vinylic cation **A** rather than the less stable bent vinylic cation **B** (Figure Si5.6). Introduction of the trimethylsilyl group, however, overrides this preference by stabilizing the silicon β carbocation **C** relative to the α carbocation **D**.

Si5.2 Vinylsilanes

Hydrosilylation and its application to the synthesis of vinylsilanes

Hydrosilylation is the term used for the addition of a silicon–hydrogen bond to unsaturated linkages such as carbonyl groups, alkenes, and alkynes (Equations Si5.2–5).

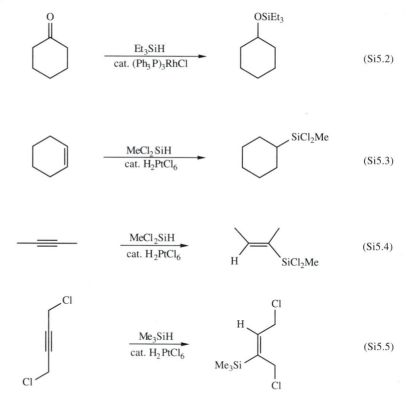

(Si5.2)

(Si5.3)

(Si5.4)

(Si5.5)

Zinc dechlorination of the product from Equation Si5.5 gives a diene which has been used in Diels–Alder reactions.

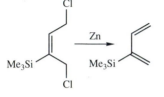

The reaction is catalysed by many transition-metal complexes, and a mechanism for the hydrosilylation of an alkene under transition-metal catalysis is depicted in Figure Si5.7. Initial coordination of the alkene to the metal is followed by *cis* addition of the silicon–hydrogen bond. A hydride migratory insertion and elimination of the product silane complete the cycle.

As illustrated in Equations Si5.4 and Si5.5, hydrosilylation of alkynes produces vinylsilanes. Chloroplatinic acid is the reagent of choice and the reaction results in *cis* addition. Good regioselectivity may result when terminal alkynes are substrates with the silyl group adding to the least hindered end of the triple bond (Equation Si5.6).

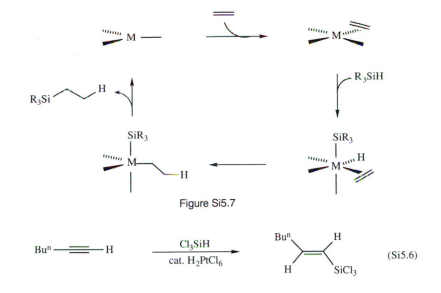

Figure Si5.7

$$Bu^n \equiv H \xrightarrow[\text{cat. } H_2PtCl_6]{Cl_3SiH} \begin{array}{c} Bu^n \quad H \\ \diagdown \diagup \\ H \quad SiCl_3 \end{array} \quad (Si5.6)$$

Other approaches to vinylsilanes

Other routes to vinylsilanes include (a) the partial hydrogenation of alkynylsilanes (Equations Si5.7 and Si5.8),

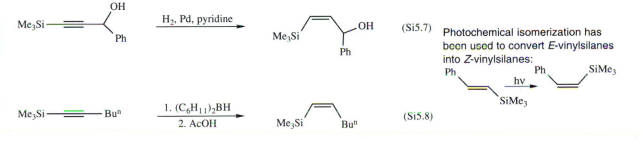

$$(Si5.7)$$

Photochemical isomerization has been used to convert *E*-vinylsilanes into *Z*-vinylsilanes:

$$(Si5.8)$$

and (b) replacement of the halogen in a vinyl halide with a metal in conjunction with an electrophilic quench (Equations Si5.9–11),

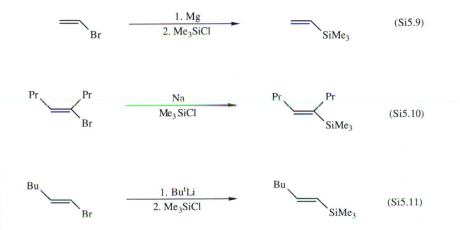

$$(Si5.9)$$

$$(Si5.10)$$

$$(Si5.11)$$

Electrophilic substitution reactions of vinylsilanes

Consider the reactions of (*E*)- and (*Z*)-β-trimethylsilylstyrene with DCl which are depicted in Figure Si5.8.

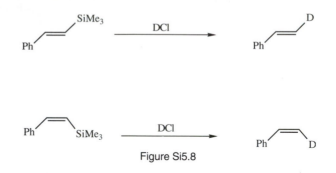

Figure Si5.8

In these reactions electrophilic substitution of the silyl group has occurred. Furthermore, note that the silyl group is replaced with retention of the stereochemistry of the alkene. This may be explained by the sequence of events depicted in Figure Si5.9. The electrophile (D^+) adds to the double bond to produce a carbocation β to the silicon atom. Rotation around the central carbon–carbon bond then occurs to bring the empty *p* orbital and the carbon–silicon bond which is to be broken into the same plane so that elimination to give the product alkene can take place. Rotation occurs in the direction which avoids placing the empty *p* orbital and the carbon–silicon bond at 90° to each other, a high-energy situation in which overlap is minimal.

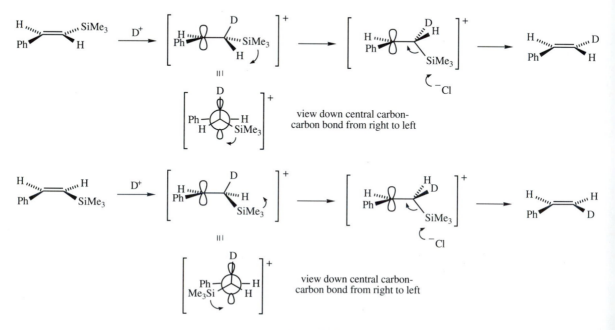

Figure Si5.9

Several further examples of electrophilic substitution reactions of vinylsilanes are depicted in Equations Si5.12–14. As in the examples given in Figure Si5.8, the geometry of the double bond is maintained in each case and this is the normal outcome of electrophilic substitution reactions of vinylsilanes.

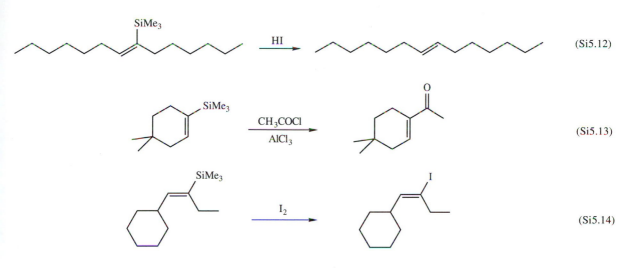

(Si5.12)

(Si5.13)

(Si5.14)

In the examples of electrophilic substitution of vinylsilanes described above, collapse of the carbon–silicon bond to form a carbon–carbon π bond is faster than nucleophilic attack on the carbocation by the anionic counterion of the electrophile. This is not always the case. For example, on treatment of (*E*)-1-trimethylsilylpropene with bromine followed by aqueous ethanol, (*Z*)-1-bromopropene is formed almost exclusively (Figure Si5.10). This is consistent with *anti* addition of bromine to the double bond followed by *anti* elimination of trimethylsilyl bromide.

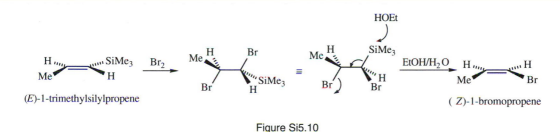

(*E*)-1-trimethylsilylpropene

(*Z*)-1-bromopropene

Figure Si5.10

Electrophilic substitution of vinylsilanes may be combined with the Nazarov cyclization to build five-membered rings, as depicted in Figures Si5.11 and Si5.12 (LA = Lewis acid). Note that the reaction depicted in Figure Si5.12 is anomalous because the vinylsilane is attacked by the electrophile at the end of the double bond remote from silicon. This is because the sulphur lone pairs stabilize a carbocation more effectively than the silicon β effect.

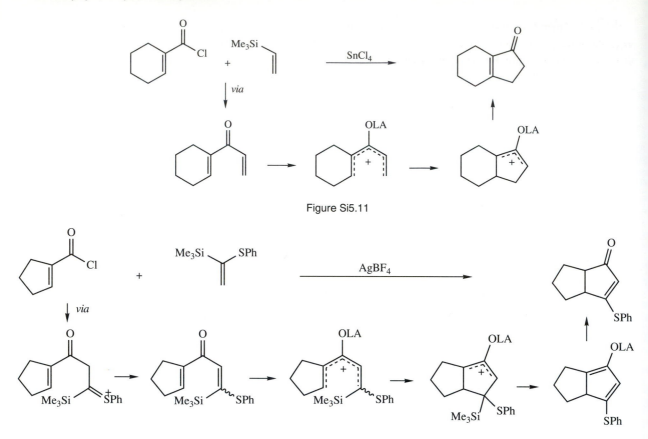

Figure Si5.11

Figure Si5.12

Nucleophilic addition to vinylsilanes

Nucleophilic addition to vinylsilanes is facilitated by the fact that it leads to the formation of an anion which is α to silicon; the anion is thus relatively stable. The reaction should be compared to the Michael addition, in which the anion generated is stabilized by an adjacent carbonyl group (Figure Si5.13).

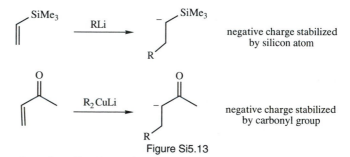

Figure Si5.13

Examples of applications of nucleophilic addition to vinylsilanes are to be found in Equation Si4.1 and Figure Si4.4.

α,β-Epoxysilanes

α,β-Epoxysilanes are readily obtained from vinylsilanes by oxidation with the peracid *m*-chloroperbenzoic acid (Equations Si5.15 and Si5.16).

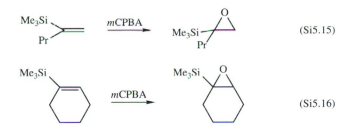

(Si5.15)

(Si5.16)

Alternatively they may be synthesized by adding α-chloro-α-lithio-α-trimethylsilanes to aldehydes or ketones (Equation Si5.17).

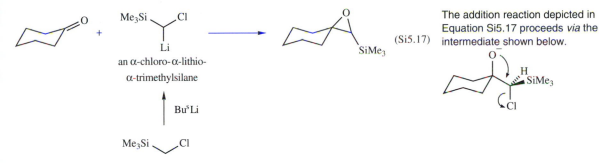

(Si5.17)

The addition reaction depicted in Equation Si5.17 proceeds *via* the intermediate shown below.

α,β-Epoxysilanes are attacked by nucleophiles at the carbon atom bearing the silyl group as illustrated in Equations Si5.18 and Si5.19. The regiospecificity observed may be accounted for by invoking stabilization by the silyl group of the build-up of negative character in an S_N2 transition state.

Recall that substitution α to silicon occurs rapidly.

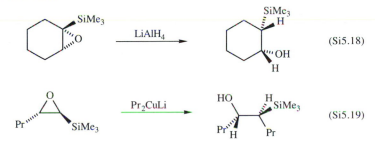

(Si5.18)

(Si5.19)

Nucleophilic addition to α,β-epoxysilanes is part of a reaction sequence which has been used to effect 1,2-transposition of carbonyl groups (Figure Si5.14).

When H_2O is added to α,β-epoxysilanes in the presence of acid, the α,β-epoxysilanes are converted into aldehydes or ketones in which the carbonyl group is formed from the carbon atom which originally carried the silyl group (Equations Si5.20 and Si5.21).

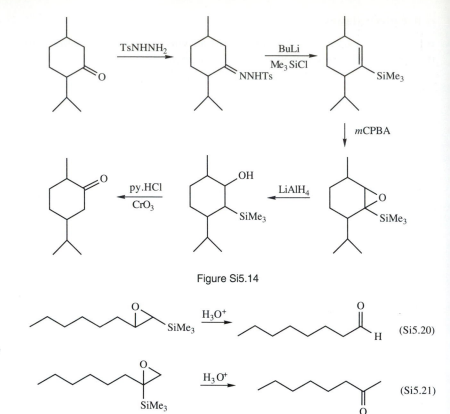

Figure Si5.14

Note that generation of an α,β-epoxysilane from an aldehyde followed by addition of water in the presence of acid constitutes the transformation of RCHO to RCH$_2$CHO.

(Si5.20)

(Si5.21)

A pathway which explains this transformation is shown in Figure Si5.15.

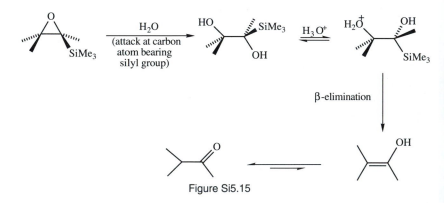

Figure Si5.15

Thus vinylsilanes may be regarded as masked carbonyl groups as exemplified by the two reaction sequences depicted in Equation Si5.22 and Si5.23. Note that in the second example the cuprate is acting as an acyl anion equivalent.

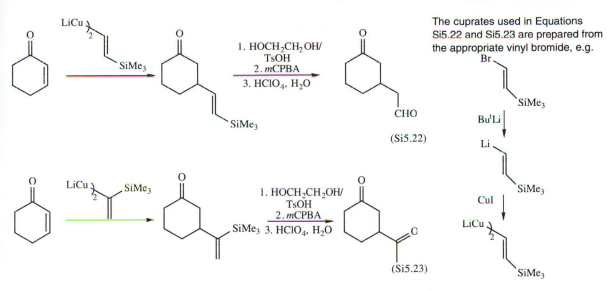

(Si5.22)

(Si5.23)

The cuprates used in Equations Si5.22 and Si5.23 are prepared from the appropriate vinyl bromide, e.g.

Finally, α,β-epoxysilanes rearrange stereoselectively under boron trifluoride catalysis to give silyl enol ethers (Equations Si5.24 and Si5.25).

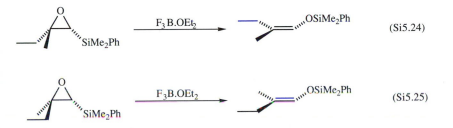

(Si5.24)

(Si5.25)

In these reactions boron trifluoride assisted ring opening occurs to give a carbocation β to the silyl group and then rearrangement gives the product silyl enol ether (Figure Si5.16 illustrates the pathway taken by the reaction depicted in Equation Si5.24.)

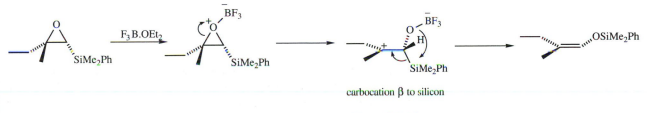

carbocation β to silicon

Figure Si5.16

Si5.3   Arylsilanes

Arylsilanes are normally prepared by quenching an aryl–metal system with a chlorosilane. The aryl–metal component is formed either by deprotonation of an activated site or by metal–halogen exchange (Equations Si5.26–29).

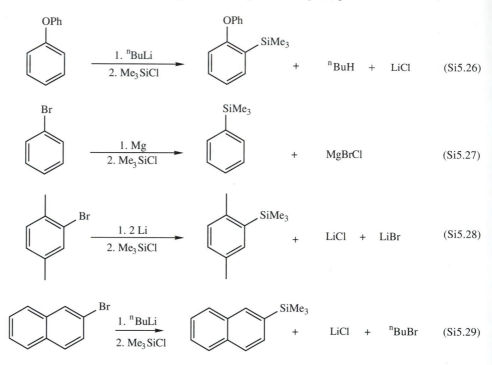

Arylsilanes undergo a reaction with electrophiles known as *ipso*-desilylation. The overall result is that the incoming electrophile replaces the silyl substituent on the aromatic nucleus (Equation Si5.30).

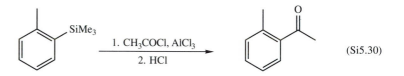

The first step of the reaction is rate-determining and forms an intermediate cationic adduct, one canonical of which is shown in Figure Si5.17. Attack at the carbon atom bearing the silyl substituent is preferred over attack at the *ortho*, *meta*, or *para* positions because the transition state preceding this intermediate is stabilized by a favourable interaction between the incipient positive charge and the carbon–silicon bond (stabilization of β carbocation effect). (Note that attack at a *meta* carbon would also lead to an intermediate bearing positive character β to the silyl group but that in this case the empty *p* orbital and the carbon–silicon bond would be orthogonal to each other.

Thus the energy of the intermediate and, more importantly, the energy of the preceding transition state would not be lowered, so this pathway is not observed.)

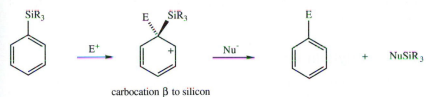

carbocation β to silicon

Figure Si5.17

Si6. Allylsilanes and acylsilanes

Si6.1 Allylsilanes

Silylation of a metal–allyl species is the most direct method available for making allylsilanes (Equation Si6.1), but several other approaches have been used including the Wittig reaction (Equation Si6.2).

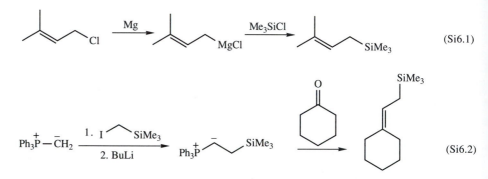

(Si6.1)

(Si6.2)

Allylsilanes react with electrophiles at their γ carbon atom as attack at this position generates a carbocation β to silicon. Displacement of the silyl group subsequently produces a new alkene (Figure Si6.1).

Recall that maximum stabilization of a β-carbocation only occurs if the vacant *p* orbital and the carbon–silicon bond can be in the same plane.

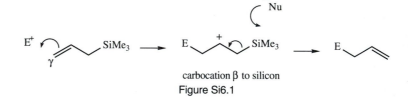

carbocation β to silicon

Figure Si6.1

Equations Si6.3–8 depict reactions of allylsilanes with various electrophiles and illustrate that allylsilanes will react with a wide range of electron-deficient species.

(Si6.3)

Lewis acid assisted addition to a ketone

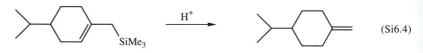

(Si6.4)

addition to a proton

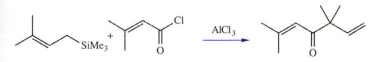

Lewis acid assisted addition to an acid chloride

(Si6.5)

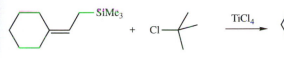

Lewis acid assisted addition to an alkyl halide

(Si6.6)

The regiochemical control observed in reactions Si6.4–6 should be contrasted with the regiochemical problems often associated with adding, for example, unsymmetrical allyl Grignard reagents to electrophiles.

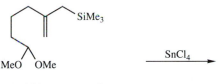

Lewis acid assisted addition to an α,β-unsaturated ketone

(Si6.7)

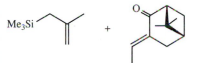

Lewis acid assisted addition to an acetal

(Si6.8)

Allylsilanes generated from γ-hydroxyvinylsilanes and enol ethers *via* a Claisen rearrangement have been used to synthesize cyclopenten-4-ols (Figure Si6.2).

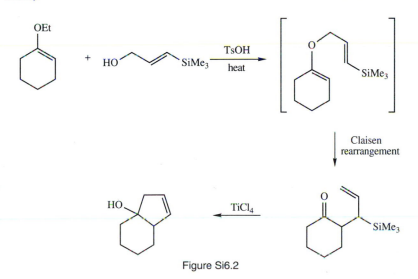

Figure Si6.2

Addition of an allylsilane to acid chlorides has been used to synthesize ipsenol and ipsdienol, principal components of the aggregation pheromone of the bark beetle *Ips paraconfusus* (Figure Si6.3). It is of note that the allylsilane used in these syntheses is a versatile source of the isoprene unit.

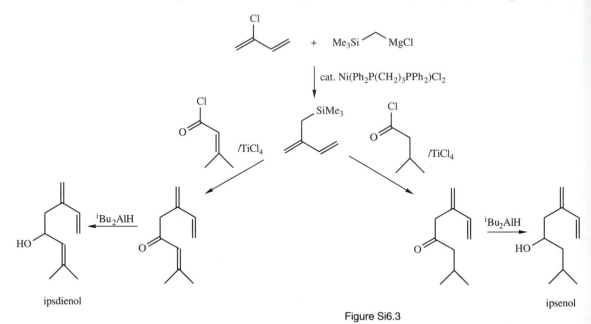

Figure Si6.3

Allylsilanes are readily deprotonated as the anion generated is stabilized not only by conjugation with the adjacent double bond but also by the neighbouring silyl group. The anion may react with electrophiles through either its α-carbon atom or its γ-carbon atom. The regiochemical and stereochemical outcome of these reactions depends on several factors of which the most important is probably the identity of the counterion (Equations Si6.9–12).

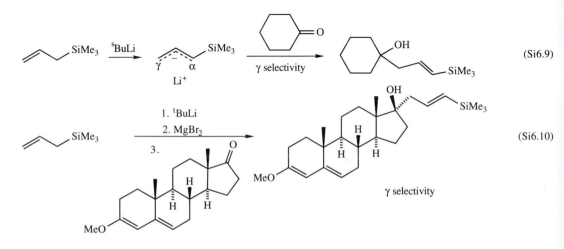

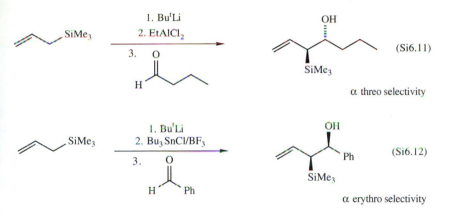

(Si6.11)

α threo selectivity

(Si6.12)

α erythro selectivity

Si6.2 Acylsilanes

One of the most direct methods of synthesizing acylsilanes involves deprotonation and silylation of the 1,3-dithiane derivative of an aldehyde. Unmasking the carbonyl group reveals the acylsilane (Figure Si6.4).

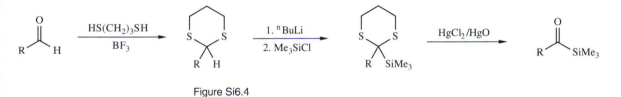

Figure Si6.4

Acylsilanes have been synthesized by several other approaches including (i) a reverse Brook rearrangement (silyl-Wittig rearrangement) of allyltrimethylsilyl ether followed by oxidation (Figure Si6.5)

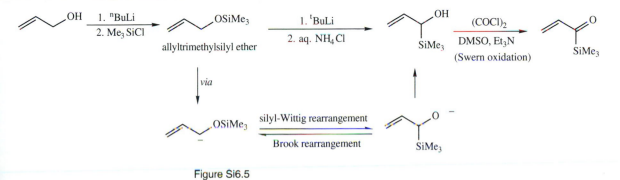

Figure Si6.5

and (ii) reductive cleavage of carbon–sulphur bonds in the silylenol ethers derived from thiol esters (Figure Si6.6).

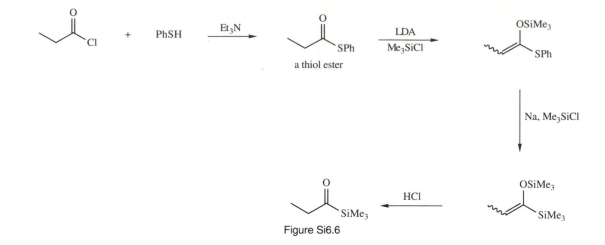

Figure Si6.6

Although acylsilanes react like ordinary ketones with some reagents (Equations Si.6.13–15),

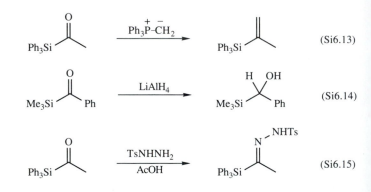

(Si6.13)

(Si6.14)

(Si6.15)

much of their reactivity is quite different. It is dominated by the tendency of the intermediates derived from nucleophilic attack to undergo the Brook rearrangement and many of the reactions of acylsilanes may be described by the general scheme shown in Figure Si6.7.

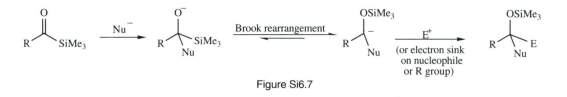

Figure Si6.7

For example, addition of the anion of phenyl ethyl sulphone to benzyl trimethylsilyl ketone gives a silyl enol ether by the pathway depicted in Figure Si6.8.

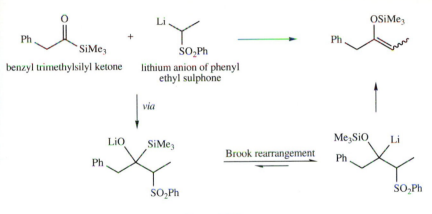

benzyl trimethylsilyl ketone   lithium anion of phenyl ethyl sulphone

Figure Si6.8

Related reactions have been used to generate regiochemically complementary silyl enol ethers of 1-phenylpentan-3-one (Figure Si6.9).

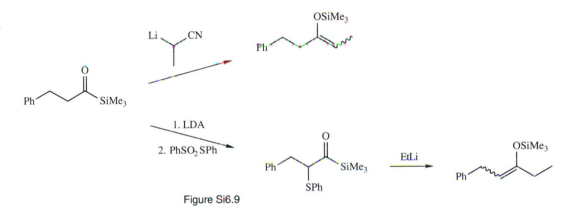

Figure Si6.9

When lithium acetylides are added to acylsilanes and the reaction mixture is quenched with an electrophile, allenol silyl ethers are formed (Figure Si6.10).

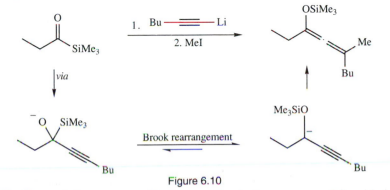

Figure 6.10

Finally, acylsilanes react with electrophiles in the presence of fluoride ion to give simple ketones (Figure Si6.11).

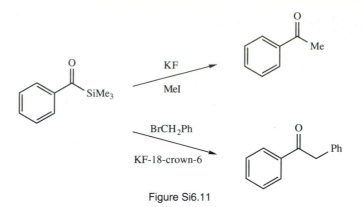

Figure Si6.11

The reaction is probably initiated by fluoride attack on the carbonyl group, as shown in Figure Si6.12. A Brook rearrangement, anion quenching, and fluoride-induced collapse of the tetrahedral intermediate then complete a plausible pathway for this transformation.

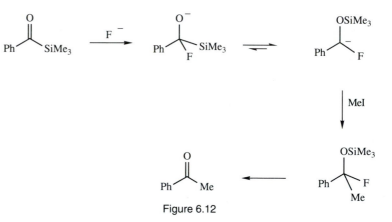

Figure 6.12

Further reading

Colvin, E.W. (1981). *Silicon in organic synthesis.* Butterworths, London. [Revised edition, Krieger Press, Florida, 1985.]

Magnus, P.D., Sarkar, T., and Djuric, S. (1982). Organosilicon compounds in organic synthesis. In *Comprehensive organometallic chemistry*, (ed. G. Wilkinson, F.G.A. Stone, and E.W. Abel), Vol. 7, pp. 515–659. Pergamon Press, Oxford.

Fleming, I. (1979) Organic silicon chemistry. In *Comprehensive organic chemistry*, (ed. D.H.R. Barton and W.D. Ollis), Vol. 3, pp. 541–659. Pergamon Press, Oxford.

Colvin, E.W. (1988) Silicon reagents in organic synthesis. Academic Press, London.

Brook, A.G. (1974). Some molecular rearrangements of organosilicon compounds. *Accounts of Chemical. Re*search, **7**, 77–84.

Lalonde, M. and Chan, T.H. (1985). Use of organosilicon reagents as protective groups in organic synthesis. *Synthesis*, 817–845.

Brownbridge, P. (1983). Silyl enol ethers in synthesis. *Synthesis*, 1–28 and 85–104.

Ager, D.J. (1984). The Peterson reaction. *Synthesis*, 384–398.

Ricci, A. and Degl'Innocenti, A. (1989). Synthesis and synthetic potential of acylsilanes. *Synthesis*, 647–660.

Bulman Page, P.C., Klair, S.S., and Rosenthal, S. (1990). Synthesis and chemistry of acylsilanes. *Chemical Society Re*views, **19**, 147–195.

Boron

alcohol synthesis 10–12, 21, 22, 23, 31–5, 43
aldehyde synthesis 18–19
alkane synthesis 13
alkene synthesis 14, 17–18, 25–7
alkyne synthesis 27–8
B-allyl-9-BBN 32
allylboranes 31–5
amine synthesis 15–16
asymmetric hydroboration 7–8, 11–12

9-BBN 4–5
9-BBNOTf 36
9-borabicyclo[3.3.1]nonane 4–5
borane 1, 3, 4, 5
boron enolates 35–41
boronic esters 42–5
Bu$_2$BOTf 36

catecholborane 14
the cyanoborate process 21–2

diborane 1
dibromoborane 27
dicyclohexylborane 3
diene synthesis 29–30
diisopinocampheylborane 6, 7–8, 12
dilongifolylborane 5–6
disiamylborane 3
diyne synthesis 28

enyne synthesis 28–9

haloalkane synthesis 14–15
homochrial alkylboranes 5–6

IpcBH$_2$ 6, 11
Ipc$_2$BH 6, 7–8, 12

ketone synthesis 13, 19–20, 22, 23

Lgf$_2$BH 5–6

monoisopinocampheylborane 6, 11

organoboranes
    amination 15–16
    carbonylation 18–21
    cyanidation 21–2
    halogenolysis 14–15
    isomerization 17–18
    oxidation 9–13
    protonolysis 13–14
    reaction with dichloromethyl methyl ether 22–3
    reaction with α-halocarbonyl compounds 24

Sia$_2$BH 3

thexylborane 3–4
triphenylborane 28

Silicon

acyloin condensation 64–6
acylsilanes 87–90
alkynylsilanes 71–4, 75
allenol silyl ethers 89
allylsilanes 84–7
arylsilanes 82–3

bond lengths 47
bond strengths 47
Brook rearrangement 49–50, 87–90
t-butyldimethylchlorosilane 52
t-butyldiphenylchlorosilane 52

Claisen rearrangement 62–4, 85

Danishefsky's diene 59
ipso-desilylation 82–3
Diels–Alder reaction 58–62

α, β-epoxy silanes 67, 79–81

hydrosilylation 56, 74–5
β-hydroxysilanes 67–70

β-ketosilanes 68

Nazarov cyclization 77–8

Peterson olefination 67–70

1,2-bis-siloxy alkenes 64–6
silyl enol ethers
    formation 55–6
    reactions 56–8
silyl ketene acetals 62–4
S$_N$2-Si mechanism 48

TBDMSCl 52
TBDPSCl 52
TIPSCl 52
TMSDEA 51
TMSI 51
triisopropylchlorosilane 52
trimethylchlorosilane 51
trimethylsilyl chloride 51
N-trimethylsilyldiethylamine 51
N-trimethylsilylimidazole 51
2-trimethylsilyloxybuta-1,3-dienes 58–62

vinylsilanes 74–8